SOCIAL INTERACTION AND ORGANISATIONAL CHANGE

Series on Technology Management

SERIES ON TECHNOLOGY MANAGEMENT – VOL. 6

SOCIAL INTERACTION AND ORGANISATIONAL CHANGE

ASTON PERSPECTIVES ON INNOVATION NETWORKS

EDITORS

OSWALD JONES, STEVE CONWAY & FRED STEWARD

Aston Business School, Aston University, UK

Imperial College Press

ICP

Published by

Imperial College Press
57 Shelton Street
Covent Garden
London WC2H 9HE

Distributed by

World Scientific Publishing Co. Pte. Ltd.
P O Box 128, Farrer Road, Singapore 912805
USA office: Suite 1B, 1060 Main Street, River Edge, NJ 07661
UK office: 57 Shelton Street, Covent Garden, London WC2H 9HE

British Library Cataloguing-in-Publication Data
A catalogue record for this book is available from the British Library.

SOCIAL INTERACTION AND ORGANISATIONAL CHANGE
Aston Perspectives on Innovation Networks

ISBN 1-86094-203-2

Printed in Singapore by Uto-Print

Preface

It is widely acknowledged in the innovation literature that networks are central to the identification, acquisition and development of new technologies. This book follows work by those such as Granovetter who argue that all activity is embedded in complex networks of social relations including: family, state, educational and professional background, religion, gender and ethnicity. It is therefore important to acknowledge that managerial decision-making occurs within a context of personal and organisational networks. Such relationships are sometimes analysed in terms of what Williamson describes as the 'transaction costs' incurred in the purchase of materials, components or technologies. While others argue that *trust* between individual employees and a firm's customers and suppliers is an essential feature of business relationships.

In the book we bring together a number of authors associated with Aston Business School who are broadly researching the topic of innovation networks. Although there is no single *perspective*, contributors agree that networks are subjective structures that cannot be separated from their social context nor from the activities of social actors. Our objective is to illustrate the way in which innovation networks are formed and sustained in a variety of organisational settings: the public sector, public-private collaboration, national policy level, the direct action movement as well as the more traditional focus on manufacturing firms. The strength of the network approach is that, on the one hand, it allows detailed analyses of the dyadic links mobilised in the innovation process while, on the other, it provides a framework for exploring the multiple sources and pluralistic patterns of communication typical of innovation activity. In contrast to much innovation network research undertaken in recent years, our interest is concerned as much with notions of 'network as method' as with 'network as a phenomenon'.

The Contributors

At the time this book was conceived all contributors were either members of academic staff or, in other cases, completing their doctorates at Aston Business School. When the book was completed a number of the contributors had moved on to other institutions. However, the conception and execution was very much based on various 'Aston perspectives' as all authors had been attached to the organisation for some time and those that are now employed by other business schools remain part of the Aston 'network'.

Martin Beckinsale recently completed his PhD on the topic of 'Strategic Innovation Networks' in medium-sized manufacturing firms. He is now employed as a Researcher at Warwick Business School studying the adoption of internet technologies in SMEs.

Stuart Cooper is Lecturer in Finance and Accounting at Aston Business School. His interests include social and environmental accounting, stakeholder theory, performance management.

Steve Conway is Lecturer in Innovation at Aston Business School. Steve's interests include social and organisational networks in innovation, knowledge and learning, the graphical representation of networks, and the social shaping technology.

David Crowther is Reader in Marketing at The Business School, University of North London. His interests include the exploration of alternative communities, accountability, and social marketing.

Tim Edwards recently completed his PhD in which he examined collaborative alliances and innovation in small firms. Tim is currently employed as a Research Associate at Cardiff Business School working on an ERDF project considering the innovative behaviour of SMEs in industrial South Wales. His research interests include critical-realist perspectives on organisational reproduction and the sociology of innovation.

Reiner Grundmann is Senior Lecturer in Sociology at Aston Business School. Reiner's interests include global environmental problems, science and society, the EU and the public sphere.

Oswald Jones recently left Aston to join Manchester Metropolitan University as Professor of Innovation and Entrepreneurship. His research interests include innovation management in mature companies, intrapreneurship and 'green' startup companies.

David Parker is Professor of Business Economics and Strategy at Aston Business School. David's interests focus on privatisation and regulation of network industries from both an economic and management perspective.

Kirit Vaidya is Lecturer in Business Economics and Director of the Aston MSc in International Business. Kirit's research interests include technology transfer and economic and industrial development in China and East Asia.

Jim Love is Professor of Economics and International Business at Aston Business School. His research interests include the economic determinants of product innovation in manufacturing industry and the transaction-cost and resource-based theories of the firm.

Stephen Osborne is Professor in Public Management and Director of Aston Business School's doctoral programme. He is currently editor of the *Public Management: An International Journal of Research and Theory*. Stephen's research interests include innovation in the public and voluntary sectors.

Fred Steward is Reader in Innovation and Director of the Innovation Research Centre at Aston Business School. Fred's interests centre on innovation networks, discourse, strategy, and social shaping of innovation.

Contents

Chapter 1

Introduction: Social Interaction and Organisational Change

Oswald Jones, Steve Conway and Fred Steward

Introduction

This book is composed of contributions from authors belonging to the Innovation Research Centre at Aston Business School. The Research Centre has members from a variety of academic disciplines including technology studies, organisational behaviour, organisational theory, economics, sociology, accountancy and public sector management. Our aim is not to provide a definitive theory of innovation networks but rather to illustrate the complex social activity associated with the interaction of technological and market opportunities. We intend to provide an overview of what we regard as the key contributions to literature on innovation networks. In examining this literature, we classify each article according to the underlying theory, the methodological approach and nature of analysis as well as the size of the sample on which the study is based. Later in this chapter, we also summarise the nine contributions to this book which vary from transaction cost economics (Parker and Vaidya) to postmodern networks of ecoprotestors (Crowther and Cooper). In our concluding

1

chapter we draw together the main issues which emerge from our review of the literature and key points from the work of our contributors. This summary is then used to set out a research agenda for future studies on innovation networks.

It has recently been pointed out that there has been an impressive accumulation of studies focusing on organizational relations and networks over the last ten years (Oliver and Ebers, 1998:549; Grandori and Soda, 1995). At the same time, the authors suggest that this work has not resulted in an accumulation of knowledge nor of conceptual consolidation. By examining 158 articles on networks from four leading journals, Oliver and Ebers posit that there are three core theoretical approaches or 'structural beacons' which provide orientation: resource dependency (Pfeffer and Salancik, 1978), political power (Zald, 1970) and 'network approaches' (Powell, 1990; Burt, 1992). It is claimed that this research demonstrates a 'central paradigm' which 'has tended to view inter-organizational networking as a response to dependencies among organizations that aims at enhancing the power and control of the networking organizations in order to foster their success' (Oliver and Ebers, 1998:565). Methodological approaches are dominated by cross-sectional, quantitative empirical studies carried out at the organisational level. Interestingly, despite this adherence to positivistic approaches 'there seems to be no clear consensus in the field on which outcomes are of central interest to research' (Oliver and Ebers, 1998:566). Perhaps not surprisingly given the positivistic orientation of most studies, the authors point out that trust and opportunism seem to be 'less important...than one might have assumed'. However, they do acknowledge that qualitative research, which is more likely to reveal social phenomena such as interpersonal trust, is 'under-represented in the field'.

In the same special issue of *Organization Studies*, Sobrero and Schrader (1998) surveyed 40 journals in an attempt to develop a 'metatheory' of inter-firm relationships. Despite many network studies which vary from industrial districts (Piore and Sabel, 1984) to detailed microsociological approaches (Granovetter, 1985; Steward and

Conway, 1996) there has been little attention devoted to analysing 'the detailed structuring of those relationships' (Sobrero and Schrader, 1998). In an attempt to resolve this problem, the authors suggest that there are two dimensions which are 'fundamental' to the management of inter-firm relationships: contractual and procedural coordination. Contractual coordination refers to the legally defined exchange of rights (Stinchcombe, 1990; Williamson, 1985) while procedural coordination refers to the structural mechanisms which are necessary for the exchange of information and organisational learning (Burns and Stalker, 1961; Duncan, 1976; Levitt and March, 1988; Nonaka and Takeuchi, 1995). The separation of organisational responsibility means that senior managers and lawyers will be responsible for contractual coordination and business unit managers and functional managers (R&D) will be responsible for procedural coordination.

Sobrero and Schrader (1998:590) quote Doz *et al.* (1989:136) who state that actual coordination is achieved not through contractual means but by patterns of communication involving individual employees: 'Top management puts together strategic alliances and sets the legal parameters for exchange. But what actually gets traded is determined by day-to-day interactions of engineers, marketers, and product developers'. In other words, there is emphasis on the processual elements which underpin the exchange of information and knowledge. Coordinating activities are related to the distinction between uncertainties about the means needed to attain a particular goal and uncertainties about the goal itself (Thompson, 1967). The level of uncertainty combined with issues of 'asset specificity' has direct implications for the structuring of relationships between cooperating organisations.

In attempting to link the various perspectives, Sobrero and Schrader (1998) develop a model which has two dimensions: task characteristics (level of uncertainty) and structuring dimensions (contractual and procedural coordinating mechanisms). The authors then set out to test a number of hypotheses generated by their model: for example, 'Do task characteristics influence the level of contractual coordination and procedural coordination?' (Sobrero and Schrader, 1998:595). A survey of 40 journals revealed 118 'networking' articles which were reduced to 32 studies which combined an explicit theoretical framework and empirical data. The meta-analysis confirmed links between task characteristics and contractual coordination but evidence related to the role of procedural coordination was 'ambiguous'. By way of explanation the authors point to problems with operationalising certain constructs in some studies as well as greater emphasis on contractual coordination. However, evidence from the paper does have implications for inter-organisational research: first, the unit of analysis should be a single relationship, secondly a range of relationships should be examined (not only joint ventures and strategic alliances). Thirdly, the 'relational perspective' is a 'promising and rather under-explored area of development in inter-organizational research' (Sobrero and Schrader, 1998:609). That is, there is a need for work in the social network tradition to be applied to the study of links between firms.

Actors and Innovation Networks

Hakansson's (1987) book which examined industrial technological development signified a growing interest in the study of networks. Recently, a number of other edited books reflect the importance of networks both as a new organisational form and a way in which management complement their internal competences (Nohria and Eccles, 1992; Coombs *et al.*, 1996; Ebers, 1997; Child and Faulkner, 1998). For two main reasons we have not reviewed either these books as a whole nor individual contributions. First, other than Coombs *et al.* (1996) the focus is more general than our own concern with 'innovation networks'. Secondly, in the UK at least, Research Assessment Exercise (RAE) pressures mean that significant contributions to the literature will almost certainly be published in journals. In other words, chapters which appear in edited books are likely to be variations on themes published elsewhere. Obviously we believe that books such as this still have a role to play in the dissemination of knowledge. Edited editions provide authors with more latitude to explore important topics than is generally the case with journals in which editors (and referees) have a narrower focus on what constitutes 'publishable articles'.

In this introduction, we examine significant contributions to the study of innovation networks published in a range of journals over the last 30 years. Unlike the two articles in the 'inter-firm' special issue of *Organization Studies* (Oliver and Ebers, 1998; Sobrero and Schrader, 1998) we do not utilise statistical techniques as a means of analysing the 49 contributions we have identified. Rather, we adopt a broadly interpretive approach in attempting to set out the key themes and theories which inform work on the topic of innovation networks. Our focus is more specific than general inter-firm relations and we concentrate our survey on ten journals in which the emphasis is technological innovation. A number of papers which address the topic of innovation networks have been published in other social science journals and we discuss the work of Granovetter (1973; 1985); Perry (1993) and Powell *et al.*, (1996).

In recent years, writers from a range of disciplinary areas have adopted or utilised the term innovation network. Contributors to regional networks, policy networks and supply chain linkages have informed and enriched our own understanding of innovation networks. Most importantly, actor-network theorists such as Callon (1986) and Latour (1988) have shifted attention from traditional 'structural' approaches to the study of social phenomenon by rejecting technological and social determinism. Technical creation (innovation) occurs as a result of numerous interactions involving researchers, scientists, technologists, engineers, managers, customers and users. Techno-economic-networks (TENs) are defined as: 'A coordinated set of heterogenous actors which included public laboratories, research centres, companies, financial institutions, government, and users' (Callon, 1992:73). TENs may be stable, producing technologies that are relatively easy to characterise, or dynamic (having significant degrees of freedom), developing unexpected technologies. Such networks are organised around three 'poles': the scientific pole which produces empirical knowledge (universities and research centres); the technical pole which develops artifacts from the empirical knowledge leading to models, pilot projects, prototypes; finally, the market pole which incorporates users who have a need which is either overt or latent.

Callon (1992) brings together the economic and the social with his concept of intermediaries which link various poles in the network: *text* (reports, journals and software), *technical objects* (telephones, fax machines, computers, vehicles), *skills* (ability to mobilise a social network as well as the technical skill required to use a computer) and finally *money* (research grants, profits, budgets, sales income and other revenues) [see Chapter 7, Tim Edwards]. The term 'sociology of translation' is used by Callon to describe the development and stabilisation of technology while reconciling the distinction between the social and the technical. In his view, networks comprise links between various 'actants' (human and non-human) which he describes as 'heterogenous entities'. Networks are configured by 'enrolling' actants through a process of negotiation which has four stages. First, *problematisation* occurs when one or more actors attempt to define a problem and set the parameters for potential solutions. Secondly, *interessment* is the stage at which the 'enrollers' attempt to persuade others that their's is the best solution. Thirdly, actors (human and non-human) are *enrolled* into the new network by coercion, seduction or consent. The final stage, *mobilisation*, occurs when rules are established to sustain the network which may represent a larger constituency (see Grint and Woolgar, 1998).

The more pragmatic 'Stockholm School' (Clark and Staunton, 1993) utilise network concepts such as actors, resources and activities to revise neo-classical economic theory. Studies of Swedish innovation networks by those such as Hakansson (1987) suggest that different time frames make synchronisation difficult and the heterogeneous nature of organisations means that 'learning' is subject to high levels of trial and error (Clark and Staunton, 1993:167). Hakasson (1987) himself is critical of work in which product development (innovation) is seen as being initiated by producers, users or some 'interplay' between the two. His view is that product development is the outcome of many small and independent events: 'The network is a social construction and as such built upon social relationships between actors' (Hakansson, 1987:493). While this view is complementary to our own approach, Hakansson and his colleagues concentrate on supply chain relationships which are common in industrial markets. For example, he concludes that an organisation's ability to network is defined by a combination of resources and activities. Similar work has been carried out by those associated with the concept of 'lean manufacturing'. Lamming (1993:245) suggests that the switch from traditional adversarial relationships between customer and supplier in the auto-industry emphasise the 'importance of joint development of new technologies, using complementary assets in the process'.

Much of the literature associated with regional networks has used the development of high-technology sectors in California as a key reference point. Although Saxenian's (1991:424) contribution to the Montreal Conference (special issue of *Research Policy*) was ostensibly about production networks she states: 'Silicon Valley demonstrates how inter-firm networks spread the costs and risks of developing new technologies and foster reciprocal innovation amongst specialist firms'. More recently, Koschatzky (1998:385) notes that studies carried out in the US using patent data or the Small Business Administration census 'reveal proximity effects in the innovation activities of industrial firms, universities and business services'. The identification of high-tech regions has implications for policy-making as national governments have tried to replicate the successes of Silicon Valley, Emilia-Romana and Baden-Wurttemberg. Porter (1998:xxiii) has been particularly influential suggesting that there is 'mutual dependence' between government and business because 'many of a company's competitive advantages lie outside the firm and are rooted in locations and industry clusters'. In recent years, the study of policy networks has emerged as an important theme in literature related to the social risks associated with innovation. Key writers such as Brandes *et al.* (1999) have adopted an approach which is similar to social network analysis for identifying relational configurations in the formation of policy networks.

Structure, Embeddedness and Innovation Networks

Social network analysis (SNA), incorporating work on scientific, R&D and entrepreneurial networks, provides tools and techniques for the collection, analysis and presentation of relational data. The approach reveals structural properties including size (number of actors) and density (number of linkages) as well as identifying cliques, key actors and isolates. SNA has largely been applied to networks of individual actors but this does not preclude the study of organisational linkages. For example, Hagedoorn and Schakenraad (1992) employ multi-dimensional scaling (MDS) techniques to reveal cliques and clusters amongst networks of organisations in a range of IT-based sectors (also see Hagedoorn and Schakenraad, 1991) .

Early work on the social organisation of science focused on defining the principal norms governing scientific activity (see Barber, 1952). Emphasis then shifted towards communication patterns of scientists and the associated social structure of scientific networks (Price, 1963; Price and Beaver, 1966; Garvey *et al.*, 1970; Lin *et al.*, 1970). By the early 1970s, researchers were linking the growth of scientific knowledge (Kuhn, 1962) with the social organisation of scientists (Crane, 1972; Griffith and Mullins, 1972). Studies concerning the communication patterns of scientists have repeatedly demonstrated the crucial role of informal personal networks in scientific information systems (Herner, 1954; Menzel, 1962; Wolek and Griffith, 1974). Research also revealed two important subgroups: *solidarity groups* (Crane, 1972) which form around highly productive scientists and *invisible colleges* (Price and Beaver, 1966; Crane, 1972) which represent elites of highly productive scientists engaged in the informal transmission of information.

The social organisation and communication patterns of engineers and technologists also became a significant area of research interest in the early 1970s (Allen, 1970; 1977; Frost and Whitley, 1971). While emergent informal communication networks were recognised as being important in engineering-based R&D it was argued that:

> "...communication patterns in the two areas of activity [science and technology] are not only largely independent of one another, but qualitatively different in their nature" (Marquis and Allen, 1966:1052).

Variations in the social organisation of scientists and engineers can in part be explained by different sets of norms and values. Allen (1977:40) argues that unlike scientists, engineers 'are limited in forming invisible colleges by the imposition of organizational barriers'. This 'enforced localism' means that engineers only work on problems that are of interest to their employer and must refrain from early disclosure of research results: 'Both of these constraints violate the rather strong scientific norms that underlie and form the basis of the invisible college' (Allen, 1977:41). Nevertheless, Larson and Rogers (1984:79) do stress the importance of friendship, job mobility, social foci and spatial proximity to the 'free-wheeling information exchange' between engineers in competing micro-electronics companies in Silicon Valley.

Johannisson and Peterson (1984:1) note the apparent paradox that, on one hand, entrepreneurship 'personifies individualism and in-dependence' while on the other hand individuals are 'very dependent on ties of trust and cooperation'. Competent entrepreneurs draw on personal networks to extend strategic competences and help resolve acute operating problems by supplementing internal resources (Birley *et al.*, 1991; Conway, 1997; Conway and Shaw, 1999). Leonard-Barton (1984:113) suggests that 'entrepreneurs who, for geographic, cultural or social reasons, lack access to *free* information through personal networks, operate with less capital than do their well-connected peers'. Equally, it is recognised that effective personal networks 'must become as complex and as heterogeneous as the daily activities of the venture' (Johannisson and Peterson, 1984:4). Inherent in the maintenance of such networks is the need for entrepreneurs to continually create *weak-ties* to prevent a few strong-ties from closing their network to opportunities and alternatives (Leonard-Barton, 1984; Aldrich and Zimmer, 1986). A recent comparative study of entrepreneurship established the importance of managerial education in encouraging the creation of networks. A lack of experience in higher education means that entrepreneurs have neither the personal contacts which are a source of information nor any real understanding of the expertise available through links with universities (Jones *et al.*, 1997). Much of this work is based on Granovetter's (1973) concept of weak ties which are an important potential source of knowledge and information (see Fletcher, 1998; Shaw, 1998). Strong ties constrain access to innovatory ideas and knowledge sources whereas weak ties open up networks which provide access to new areas of expertise. At the same time, entrepreneurial activity is embedded in complex networks of social relations which are based on family, state, educational and professional background, religion, gender and ethnicity (Granovetter, 1985).

Social Interaction and Innovation Networks

Those writing about networks generally adopt a functionalist perspective which is based on a biological metaphor which emphasises that the structure and functioning of social systems is evolutionary and adaptive (Morgan, 1986). Network scholars can be broadly divided into those who favour structural explanations (Allen, 1970; Hakansson, 1987; Powell *et al.*, 1996) in which the social system as a whole (the network) is pre-eminent over individuals and, secondly, those who consider human action (Leonard-Barton, 1984; Conway, 1994; Shaw, 1998) to be the key explanatory factor in the formation of networks. All social structures (networks) encompass the enduring patterns of social interrelations such as class as well as roles, rules and social institutions. Therefore, it is necessary to resolve the endemic agency-structure dualism by emphasising the way in which knowledgeable, reflexive social actors draw on rules and resources in their day-to-day social activity (Giddens, 1984). This recursive activity constantly recreates the 'structural properties' that provides the framework for everyday social practices (Jones, 1997a; Jones *et al.*, 2000). Giddens (1984) uses the terms system and structuration to emphasise that the rules and resources drawn on in the (re)production of social action are at the same time the means of system reproduction. Social systems do not have structures but exhibit structural properties. In structuration theory, the relationship between agency and structure is similar to the association between grammar and speech. The rules of grammar are utilised by social actors in their patterns of communication which in turn constantly recreate the structure of language.

Structuration theory and networks are brought together in Barley's (1990) analysis of new technology and the revised structural arrangements in two hospitals. Using the concepts of relational and non-relational roles he 'explicitly articulates how skills, tasks, and activities influence role relations and how role relations, in turn, affect an organization's and an occupation's structure' (Barley, 1990:98). In a recent paper, Barley and Tolbert (1997) develop a more dynamic model in which they categorise day-to-day interactions by identifying the 'scripts' used by actors. Scripts are the 'observable, recurrent activities and patterns of interaction characteristic of a particular setting' (Barley and Tolbert, 1997:98). This means that networks cannot be studied as objective social structures which are independent of human agency. As Roberts and Grabowski (1996:418) point out: 'Organisational researchers are now more explicit in acknowledging that social activity is embedded in social networks which include family, friends and co-workers as well as broader factors such as religion, gender and ethnicity'. In contrast, Latour (1999) argues that after twenty years the term 'network' has lost its freshness 'as a critical tool against notions as diverse as institutions, society, nation-state and, more generally, any flat surfaces'. He also objects to those who use actor-network in the same sense as agency-structure: 'Most of the misunderstandings about ANT have come from this coupling of terms…'. Rather, Latour sees actor-network theory as a way of paying attention to 'dissatisfactions' associated with attempts to understand society from either the micro or macro levels. Micro level (face-to-face interaction) dissatisfactions are related to the problem 'that many of the elements necessary to make sense of the situation are already in place: Norms values, culture, structure, social context'. Macro level dissatisfactions concern abstractions of terms such as structure and culture which seem too great and there is a need to reconnect to 'flesh and blood local situations from which they had started' (Latour, 1999:17). In the same volume, Law (1999:2) refers to the 'multi-national monster' of the actor-network approach now that it has been converted into 'smooth and consistent theory'.

Our interest in the topic of networks is concentrated on the way in which business organisations manage the innovation process. In other words, we are concerned with examining various stages from idea generation to the implementation of new products, services and processes. Innovation networks may involve the analysis of social interaction within organisations or the focus may be external linkages such as joint ventures and strategic alliances. In either case, the research interest could be formal linkages defined by organisational structures and contractual agreements or on the informal linkages associated with the 'grapevine' and what have been described as 'invisible colleges' (Crane, 1969). Both formal and informal linkages contribute to innovatory performance by providing access to new skills, knowledge, information and technologies. This does not mean that we adopt a managerialist approach in which networking is simply seen as a way of improving the effectiveness with which knowledge is acquired and utilised. Rather, we are attempting to better understand the process by which new products and services are created because as Schumpeter (1943) points out innovation is 'the fundamental engine that sets and keeps the capitalist engine in motion'. Therefore, it is important to question decisions about technology at organisational (Jones, 1997b) and societal levels (Steward, 1995).

In his review of the literature, Freeman (1991) discusses definitions of the term 'innovation network'. These range from the view that they are 'basic institutional arrangements to cope with systemic innovation' (Imai and Baba, 1989) to those linkages which are 'of a mainly informal and tacit nature' (Camagni, 1991). Our own definition emphasises the idea of networks as a way of understanding organisational innovation:

> An innovation network is a conceptualisation of the in-
> novation process as a complex and pluralistic pattern of
> interactions, exchanges and relationships between actors
> participating in that process.

Hence, innovation networks incorporate the formal and informal linkages which are established within and across organisational boundaries as a means of capturing codified knowledge represented in specifications, reports and software as well as tacit knowledge which can only be communicated by direct social interaction (see Nonaka and Takeuchi, 1995). Networks are formed or emerge in response to some perceived need within organisations for knowledge which will contribute towards the development of new products, processes or services. The formation of any innovation network is a process by which random patterns of social activity are gradually institutionalised into organisational routines (Nelson and Winter, 1982).

From a methodological perspective, we concur with Giddens (1984:284) who argues that 'all social research has a necessarily cultural, ethnographic or "anthropological" element to it'. This relates to what he describes as the double hermeneutic in which there is intersection between the two frames of reference represented by the world as understood by lay actors and the 'metalanguages' utilised by social scientists. That is, competent actors develop their own understanding of social life which is interpreted by social scientists making use of 'second order concepts'. Researchers must be sensitive to the complex skills utilised by actors in coordinating their everyday behaviours and practices. Structuration theory also encourages researchers to take into account the time-space constitution of social life which has traditionally been left to historians and geographers. History and context are central to studies which are intended to reveal the process of technological innovation. At the same time, Giddens (1984) rejects the search for universal laws found in natural science because they simply do not exist in the 'realm of human social conduct'.

While we have set out our own perspective on the links between social interaction and innovation we accept that there is no single theory of networks. In fact, we broadly concur with Martin and Frost (1996:610) who in discussing postmodernism make the following point: 'It is a discourse, rather than a unified theory'. At the same time, we reject an extreme relativistic perspective in which 'anything goes'. Though repudiating the idea of a metatheory of innovation networks we still believe that it is important to retain an historical perspective on organisational change. For example, the idea of the 'virtual' or networked firm is not a feature of postmodern society — such firms were typical of industrial organisation well into the C20th (Pollard, 1965). Collaborative agreements are seen by those from an economic perspective as a way of decreasing uncertainty and the likelihood of cheating. But attempts to reduce the complexity of human behaviour to mutual expectations about future behaviours are clearly inadequate. Sociological approaches to inter-organisational relations are important because *trust* is conceptualised as 'a communicative, sense-making process' (Clegg and Hardy, 1996:679). There is also a need for a more critical analysis because collaboration is often seen as an activity from which partners benefit equally. Although many collaborative deals no doubt are beneficial to all parties it is also important to recognise that such arrangements can also be exploitative: 'Existing work has been noticeably mute on questions relating to power' (Clegg and Hardy, 1996:679).

Researching Innovation Networks

Those studying innovation networks have utilised a wide range of methodologies and have adopted an eclectic array of theoretical positions. Some researchers such as Powell *et al.* (1996) adopt a positivistic approach to the study of networks based on statistical analyses of data acquired from a relatively large (225 firms) data set. In contrast, others such as Knights *et al.* (1993) adopt a post-structuralist perspective based on *data* from a single case study. In this section, we briefly summarise the contrasting approaches utilised in some of the more significant network studies undertaken in recent years.

In the past, firms organised the major elements of their R&D internally and contract research was primarily concentrated on relatively mundane activities such as toxicological tests. According to Powell *et al.* (1996) companies are increasingly performing all stages of the innovation process (discovery to marketing) through some form of networking arrangement. There are many explanations for this increase in networking including shared risk, market access, complementary assets and speed to market. Powell *et al.* (1996) focus on the argument that networking is more prevalent in sectors (biotechnology, for example) where knowledge is developing rapidly and sources of innovation are likely to be found in the 'interstices between firms' which includes universities, suppliers and customers. It is claimed that the level of technological sophistication will be 'positively correlated with the intensity and number of alliances in those sectors' (Powell *et al.*, 1996:116). The study is theoretically underpinned by a view of innovation and organisational learning as a function of the firm's access to knowledge (March, 1991; Nelson, 1990; Stinchcombe, 1990) and its 'absorptive capacity' (Cohen and Levinthal, 1990). The methodology utilised a database, assembled by the authors, of independent dedicated biotechnology firms (DBFs) and their formal contractual, inter-organizational agreements. Data were collected over a five-year period from 1990 to 1994. The authors were able to examine a range of factors associated with networking activity including heterogeneity, central connectivity, degree centrality and closeness centrality (Powell *et al.*, 1996:127–128). It was also possible to create a 'dynamic model' by tracking changes to network relationships over the five-year period. A number of important issues emerged from the study of 225 biotechnology firms, in particular, the lack of influence of traditional organisational variables such as age and firm size: the former proved 'unimportant in the context of network experience, and size was an outcome rather than a deter-minant of partnerships' (Powell *et al.*, 1996:142). A 'path-dependency' cycle of learning was also identified with early exploration leading to positive feedback which reinforced the importance of R&D alliances as the 'locus of innovation'. In addition, the 'modal' firm was typified by multiple relationships and those without ties were extremely rare.

In their article, Knights *et al.* (1993) begin by pointing out that most studies of inter-organisational networks lack any 'critical' dimension. The data are based on a study of 'Switchco' which, in 1990, was established to facilitate electronic trading between 20 insurance companies. The authors focus on those 'knowledge workers' who were responsible for establishing and developing the electronic trading network. A conceptual framework was developed by com-bining actor-network theory (Callon, 1986) and Foucault's (1986) theory of power and knowledge. Most studies emphasise 'reciprocity and mutual trust, where self-interest is sacrificed for the common good' while ignoring evidence which suggests that networks are 'embedded in institutional power relations that are hierarchical, competitive and instrumental' (Knights *et al.*, 1993:979). Actor-network theory helps reveal that the political activities within Switchco which made it difficult to establish a clear corporate view on the electronic trading network. Foucault's work is utilised as a way of moving beyond 'middle-range' theories which ignore 'wider structures of inequality'. Hence, strategic inter-organisational networks can be seen as a 'disciplinary technology' by which 'capitalist economies revolutionize their means of production' (Knights *et al.*, 1993:988). In other words, networking fits with what du Gay (1996) describes as 'new wave management' in which employees are encouraged to believe that they are acting as independent 'enterprising subjects' when, in fact, they are closely monitored and controlled by senior managers.

A study by Conway (1995) of 35 'commercially successful innovations' moves beyond the single case study while retaining the deeper analysis associated with qualitative research. The study, designed to investigate the role of informal boundary spanners, was informed by 'social exchange theory' (Uehara, 1990) which emphasises the reciprocity obligations which are a feature of social interaction. The concept of 'action sets' was used to describe the range of dyadic relationships associated with some specific social activity such as the innovation of a new product. Conway's (1995) sample was drawn from the winners of two prestigious UK innovation awards: The Queen's Award for Technological Achievement and the British Design Awards. Interviews were carried out with 'focal actors' associated with the 35 award winners to identify a range of relationships mobilised in the innovation process: 'The nature of the dyadic link (exchange of friendship, power, information or goods); the degree of formalisation; the degree of reciprocity or symmetry; the extent of multiplicity (single or multiple links); the strength or intensity of the relationship' (Conway, 1995:330). While the results indicated that economic outcomes were a feature of boundary spanning activity there was also support for the proposition that informal exchange behaviour is highly complex. In many cases, information was exchanged or resources obtained because of friendship ties with no expectation of financial return.

David who had come to work for me, used to work for CEGB, and he knew somebody called Greg. And Greg has gone up in CEGB. He phoned Greg and said: "Look we need to borrow a piece of equipment." And Greg said "Right OK" (Conway, 1995:337).

In conclusion, Conway argues that it is essential for policy makers and those involved with the management of innovation to acknowledge the importance of informal friendship ties if technological change is to be managed effectively. These three papers are not necessarily representative of empirical studies of innovation networks. Rather, the work of Powell *et al.* (1996), Knights *et al.* (1993) and Conway (1995) illustrate the range of theoretical and methodological approaches to the study of collaborative links during the innovation process. In the following section, we analyse 49 papers which we have identified as making a significant contribution to the literature on innovation networks.

Issues of Theory and Method

In Appendix 1, we summarise the contributions of 49 papers based on four factors: theory, method, analysis and sample. That is, we have attempted to identify the theory which underpins each paper as well as the methodological approach, the nature of the analysis and the sample on which the studies were based. In a substantial number of papers issues of theory, method, analysis and sample are what might be described as 'opaque'. This is particularly evident with regards to defining 'theory' which is often implicit rather than explicit (see Hagedoorn and Schakenraad, 1992; Rothwell and Dodgson, 1991; Lawton Smith *et al.*, 1991). The largest proportion of papers adopt a 'network' perspective which is described by Oliver and Ebers (1999:575) as focusing on how the position of actors in a network and the content of relations affect opportunities for action (Burt, 1992; Powell, 1990). Some authors combine network theory with another theory such as 'organisational learning' (Senker and Sharp, 1997; Tidd, 1993). If the rigorously defined theoretical approaches such as social exchange theory, social embeddedness, actor-network theory and social network theory are grouped under the generic title of networks then they account for more than 50% of the contributions.

The main methods of data collection were interviews, questionnaires and secondary data sources. Many authors lack precision in explaining their research approach in terms of number and length of interviews and whether respondents were interviewed on more than one occasion (Graham, 1998; Senker and Sharp, 1997). Similar issues are also apparent when examining research based on questionnaire data (Midgley *et al.*, 1992; Rothwell and Dodgson, 1991). Such problems are exacerbated when studies are based on secondary sources as authors appear unwilling, or perhaps regard it as unimportant, to specify the exact nature of their data and the way in which they were acquired (Gemser *et al.*, 1996; Robertson and Langlois, 1995).

Approaches to data analysis, as well as research methodologies, can be categorised as qualitative or quantitative. In general, those adopting quantitative approaches utilise some form of statistical analysis ranging from cluster analysis and multiple dimensional scaling to simple descriptive statistics. It would be fair to say that overall there is a lack of sophistication in the usage of statistical techniques amongst the majority of papers represented here. Qualitative approaches are both more varied and in some cases considerably more sophisticated in the nature of their analyses. Steward and Conway (1998) utilise discourse analysis to identify different approaches to green innovation in German and UK small firms. Macro and micro levels are combined by von Raesfeld Meijer (1998) who draws on socio-cognitive mapping to analyse a complex case study involving a project to rebuild Den Bosch bus and rail stations in Amsterdam. Most of those utilising qualitative methodologies adopt what might be described as a 'discursive' approach to their data analysis (reaching a conclusion *via* reasoned argument). Generally, this means using a particular theoretical framework to order and categorise data obtained from interviews or secondary sources. Perhaps the best example is the contribution by Knights *et al.* (1993) who combine actor-network theory and Foucault's concept of power-knowledge to critically analyse an attempt by UK insurance companies to establish an electronic trading system (see above). Another notable piece of work utilising this approach is Perry's (1993) historical analysis of the Manhattan Project.

Of the papers which we surveyed only eight were based on relatively large samples; Autio (1997) studied 130 firms; Baptista and Swann (1998) used the SPRU database to analyse 284 manufacturing firms; Hagedoorn and Schakenraad (1992) examined 1700 'alliances' between 210 firms; Hakansson (1990) made between one and six interviews in 123 firms; Karlsson and Olssson (1998) surveyed 279 Swedish firms; Koschatzky surveyed 2042 SMEs; Powell *et al.* (1996) made a longitudinal study of 225 biotechnology companies; Swan and Newell (1995) carried out 20 interviews and obtained a questionnaire sample of 189. Rothwell and Dodgson (1991) did include a sample of 100 UK and 80 Italian SMEs but there is a lack of clarity about when and how data on these firms were acquired. There are a number of mid-range studies which have samples greater than ten but less than 100 (Conway, 1995; Kreiner and Shultz, 1993; Lawton Smith *et al.*, 1991; Midgley *et al.*, 1992; Steward and Conway, 1998). We do not suggest that large samples equate with quality nor that the findings from very small samples are trivial. The cases studied by Graham (1998), Knights *et al.* (1993), Perry (1993) and von Raesfeld Meijer (1998) all tackle complex and significant topics. We recognise that there are strengths and weak-nesses associated with both approaches and our view is that social scientists should attempt to reconcile what Latour (1999) refers to as micro-macro 'dissatisfactions'.

In our concluding chapter, we use this analysis of 49 key innovation network papers as well as the contributory chapters as a basis of a future research agenda (see Salancik, 1995). At this stage, we can summarise the strengths and weaknesses of the network literature in the following manner. There are a number of important papers which have implications far beyond the relatively narrow field of innovation studies. In particular, the work of Knights *et al.* (1993) and Powell *et al.* (1996) which was summarised above. We do not regard the diversity in theory, method, analysis and sample size as a weakness but we do feel that all authors should be encouraged to clearly set out these key elements of their research approach.

Aston Perspectives

While the importance of existing literature on innovation networks is acknowledged by Jones and Beckinsale they argue that little attention is given to the role of micropolitics in creating and sustaining such networks. Micropolitics refers to the way in which key actors pursue their own career interests in attempting to gain influence, status and power. Such a perspective draws heavily on the work of Burns and Stalker (1961) who identified 'internal politics' as a key influence on organisational structure and innovation. It is suggested by Jones and Beckinsale that individual political activity occurs as managers pursue career objectives in competition with rivals and, on occasions, in conflict with broader organisational objectives. Data for the study are drawn for an ongoing seven-year study of an engineering firm with 750 employees. This longitudinal approach permits the authors to map the changing innovation networks as managers attempted to develop and market two radically new products. The strength of this approach is that it enables the authors to establish that networks are subjective structures which are inseparable from the social context in which they are formed. In other words, it is important to recognise that networks do not have an objective reality outside the activities of the actors involved with creating and sustaining those networks.

Key writers associated with both innovation studies and organisation theory have long acknowledged the importance of informal organisation. The emergence of relationships outside those prescribed by formal structures can aid communication and integration but can also create opposition and destabilisation. Conway draws on social network analysis which has its roots in the work of British anthropologists (Radcliffe-Brown) and German social psychologists (Kohler) to study the implementation of an IT system. The research is based on a Housing Trust which has five regional offices in the UK. In 1994, the midland region was selected to pilot the introduction of a major new IT system which would provide more effective links between activities in the Trust. Tensions were created by the emergence of informal activities associated with attempts to resolve a number of technical problems. Even though the mobilisation of this informal network did improve the system it also 'undermined the legitimate power and personal credibility of key actors'. Conway points out that as organisations are restructured into more organic forms the importance of informal networking will increase and therefore managers must be encouraged to accept the dilution of their power for the benefit of improving innovatory activity within organisations.

Economists approach innovation networks from a very different perspective than those whose work can be categorised as organisational theory or organisational behaviour. The contribution by Parker and Vaidya is based on 'recent developments in institutional and evolutionary economics'. The authors acknowledge that evolutionary economics is closer than neoclassical economics to Granovetter's work on embeddedness which underpins many of the contributions to this book. At the same time, the importance of positive externalities which emerge from structures of interaction are still regarded as a key element in economic analyses of innovation networks. While Parker and Vaidya accept that the application of game theory to networks remains largely theoretical they argue that it is potential a useful concept in explaining why organisations participate in cooperative relationships.

Love attempts to extend the transaction cost approach associated with new institutionalism by incorporating a network perspective. However, he does argue that writers such as Williamson place too much emphasis on the issue of 'opportunism' as a basis for 'non-market or quasi-market institutional structures' (networks). The inclusion of a resource-based view helps overcome the limitation of transaction cost economics in accounting for network relationships between firms. Love draws on an extensive survey of 15,000 manufacturing firms in the UK, Ireland and Germany to investigate this issue empirically. A number of important factors are identified from the survey, in particular, networking patterns vary substantially between plants in the three countries. German companies are much more likely to engage in *explicit* collaborative arrangements than firms in the UK or Ireland. In addition, German firms emphasise risk-sharing and stress the importance of engineering and technical inputs to the network. In contrast, firms in the UK and Ireland seem more concerned to access external expertise and links are dominated by marketing rather than technical functions. Drawing together the theoretical development of transaction cost analysis and the empirical data Love argues that 'other institutional arrangements can substitute for R&D in the innovation process'. In other words, networks are an important source of learning which is underestimated by conventional institutional approaches typified by Williamson's work.

The emergence of new technological trajectories are almost certain to carry some element of risk. Attempts to regulate the externalities associated with new technologies usually occur long after innovations have appeared in the marketplace. Steward examines the relationships between technological innovation and environmental risk. The term 'policy network' is used to describe the interaction between government and special interest groups. Following a number of crises in recent years such as BSE (bovine spongiform encephalophathy) and E-coli there has been increasing interest in food sector innovation. Because changes to methods of growing, processing and distributing food impinge on human health and the natural environment they are, Steward argues, of interest to a 'diverse array of social movements'. Mapping techniques are a useful analytical device for identifying key actors in both innovation and policy networks.

The UK's Teaching Company Scheme (TCS) is a key institutional mechanism for transferring knowledge from universities to industry. It is of particular importance in providing access to new technologies for the owners and managers of smaller firms. Edwards uses the concept of *Techno-Economic Networks* to investigate the *process* of technology transfer *via* the TCS. According to Callon (1992) Techno-Economic Networks are organised around three poles: the scientific pole (knowledge creators), the technical pole (knowledge users) and the market pole. The two-year study examines a TCS programme arranged between a small firm (25 employees) manufacturing computer cleaning equipment and a Business School in the west midlands. Senior managers within *Beta* instigated the programme because they wanted to implement a computer-based purchasing system. Problems associated with the day-to-day management of a small firm meant that the TCS Associate (a young graduate responsible for 'knowledge transfer') was given far greater responsibility than was originally intended. Because the study was longitudinal, Edwards is able to identify the 'moments of translation' associated with the construction, communication and exchange of knowledge during the innovation process.

An analysis of the links between networks and learning is the topic of Grundmann's chapter. The starting point is a critique of 'new institutionalism' represented by transaction cost economics (Williamson, 1975; 1985). Drawing on the work of Powell and Koput (1996) it is asserted that learning is a social process that means collaboration is not simply a way of compensating for lack of skills within an organisation. Grundmann argues that 'norms, values, tacit knowledge, trust and face to face interactions' play a key role in the formation of innovation networks and industrial districts. This 'structural embeddedness' perspective suggests that 'constrained' firms, which are part of a network, innovate more than 'autonomous' firms. While Grundmann accepts that networks can lead to 'lock-in and path-dependency' he argues that knowledge-based sectors in which there are high levels of inter-connectedness may not be susceptible to this failing. That is, flows of information and knowledge (learning) between firms stimulate innovatory activity within the population of firms. As Grundmann points out, this is an interesting and counter-intuitive finding.

A much neglected topic in both management and innovation literatures is the organisation of activities in the voluntary sector. Osborne argues that voluntary and non-profit organisations (VNPOs) have a long history of innovation which can be traced to the pioneering work of the Webbs (Webb and Webb, 1911). At the formation of the NHS in 1948 Beveridge certainly saw the voluntary sector as an important element of innovation in social welfare. More recently, the Griffiths Report (1988) encouraged the use of voluntary bodies to stimulate innovation in the provision of wider consumer choice. The study carried out by Osborne is based on an extensive questionnaire survey of a rural, urban and suburban locality. Twenty four case study sites were identified in the three localities and interviews carried out with 'organisational leaders'. The well-known work of Miles and Snow (1978) provides an analytical framework by which to categorise the approach to innovation in each of the sites. It is established that networks are an important element in the innovative capacity of VNPOs . Furthermore, this innovative capacity is heavily embedded in the relationship between a VNPOs and its environment. Such relational networks are perceived to be an important element in legitimising the activities of VNPOs and of ensuring their long-term survival. Another significant finding was that some case study sites claimed they were innovative when in fact it was impossible to substantiate such claims.

Another area largely ignored by most management scholars is the increasingly significant topic of the UK's direct action protest movement. Crowther and Cooper utilise an innovation network perspective to analyse two cases of ecoprotest: the Birmingham northern relief road and a road development in the centre of Derby. Ecoprotests depend on the establishment of permanent sites but at the same time are virtual organisations because of fluctuating membership which involves a variety of groups including ecoprotestors, travellers and representatives of local communities. According to the authors, the ecoprotest movement is largely self-organising and as such 'is an exemplar of a truly postmodern network'. Hence, the creation of 'econetworks' have some significant lessons for those operating in more traditional organisations. For example, such aspects as informality, lack of hierarchy, consensual decision-making and 'positive' relations within the network all contribute towards organisational innovation. Furthermore, the authors claim 'such postmodern networks are able to react to an unstable environment more quickly than conventional organisations and adapt in an evolutionary manner to circumstances at the time'.

In the final chapter, Conway *et al.* address the issue of whether innovation networks are simply a fad. Certainly widespread adoption of the networking term does suggest it is no more than the managerial 'flavour-of-the-month'. At the same time, there has been increasing interest in networks from a policy perspective which may indicate a more substantive role in promoting industrial regeneration through revitalisation of the small firm sector. From an academic perspective social network analysis has been widely used in studies of scientific activity since the 1950s (Barber, 1952). Despite this long intellectual tradition there are weaknesses in the way in which the study of innovation networks are actually carried out. The main areas of concern include an under-emphasis on data gathering, few attempts to deal with the 'network content' (flows of information), too much 'theory-less' research as well as a need to be clearer about how and why network boundaries are established. Echoing Salancik (1995), Conway *et al.* conclude that the greatest challenge remains the development of a useful and robust network theory.

Appendix 1. Contributions to innovation networks.

Author(S)	Theory	Method	Analysis	Sample
Allen (1971)	Networks and gatekeepers	Questionnaire	Statistical and mapping	R&D laboratory
Autio (1997)	Resource-based	Comparison of three databases	Statistical (t-tests)	82 (US); 29 (Fin); 19 (UK)
Baptista and Swann (1998)	New growth externalities (networks)	SPRU database; employment data (CSO)	Statistics (OLS and linear exponential modelling)	284 manufacturing firms (1975–1982)
Belussi and Arcangeli (1998)	Evolutionary	Secondary sources	Discursive/ comparative	Case studies of three Italian regions
Chesbrough and Teece (1996)	Innovation type (systemic/ autonomous)	Secondary sources	Discursive	Case study: IBM and the PC
Conway (1995)	Social exchange	Semi-structured interviews	Identifying exchange value	35 award winning firms
Conway and Steward (1998)	Social exchange	Semi-structured interviews	Graphical network maps (software)	Three SME cases
Cooke (1996)	Networks	Secondary sources	Cross-national comparisons	Seven regions and/or countries
Cooke and Morgan (1994)	Public policy	Interviews and secondary sources	Basic statistics; structured analysis	80 respondents (gov, TU, ind)
Czpiel (1975)	Sociological diffusion	Interviews: technicians and managers	Sociometric mapping	18 continuous casting firms
DeBresson and Amesse (1991)	Networks	Literature review		
de Laat (1999)	Innovation type (systemic/ autonomous)	Secondary sources (trade journals)	Discursive	Case study: digital video disc (DVD)
Fletcher (1998)	Social embeddedness	Ethnographic	Attributing values and meaning	SME (70 employees)
Foray (1991)	Transaction costs	Secondary sources	Comparative	Two case studies
Freeman (1991)	Networks	Literature review		

Appendix 1 (*Continued*)

Author(S)	Theory	Method	Analysis	Sample
Frost and Whitley (1971)	Networks and gatekeepers	Two separate questionnaires	Statistics (Mann-Whitney)	R&D laboratory
Gemser *et al.* (1971)	Life-cycle and appropriability	Secondary sources	Comparative	Three industry case studies
Graham (1998)	Actor network theory	Extensive interviews	Discursive	UK electronic livestock auctions
Hagedoorn and Schakenraad (1992)	Networks	CATI databank	Cluster analysis; multi-dimensional scaling	210 firms — 1700 alliances
Hakansson (1990)	Networks	Structured interviews — 27 items	Descriptive statistics	123 firms (one to six interviews)
Hislop *et al.* (1997)	Articulation process	Literature review		
Hobday (1994)	Complementary assets and PLC	Secondary data	Discursive	Silicon Valley in 1980s
Jones *et al.* (1998)	Structuration theory	Literature review		
Karlsson and Olsson (1998)	Network externalities	Postal survey (questionnaire)	Statistical (tobit) comparative	140 large firms; 130 SMEs
Knights *et al.* (1993)	Actor network theory + Foucault	Qualitative (not stated)	Discursive	Switchco
Koschatzky (1998)	Spillovers and networks	Survey — questionnaire	Logistic regression	2042 SMEs in four regions
Koski (1999)	Network externalities	Questionnaire — input-output data	Statistics (OLS and descriptive)	Finnish firms: 15 (1994); and six (1996)
Kreiner and Shultz (1993)	Barter economy	16 in-depth interviews	Discursive	Biotechnology research directors
Laamanen and Autio (1996)	Life-cycle and dyn. Comple-mentarities	Secondary data	Discursive	Finnish electronics industry
Langlois and Robertson (1992)	Networks	Secondary sources	Comparative	Case studies: hi-fi stereo and PCs

Appendix 1 *(Continued)*

Lawton Smith *et al.* (1991)	Life-cycle and dyn. comple-mentarities	Secondary data	Discursive	Finish electronics industry
Midgley *et al.* (1992)	Structural equivalence	Questionnaire	Network mapping	16 Australian firms
Perry (1993)	Institutionalist	Secondary sources	Retrospective mapping	Case study: Manhattan project
Powell *et al.* (1996)	Networks and learning	Analysis of formal agreements	Multiple regression and correlation	225 firms (largely US)
Robertson and Langlois (1995)	Networks	Secondary sources	Discursive	None
Robertson *et al.* (1996)	Innovation diffusion	Semi-structured interviews + documents	Structured comparisons	Three large firms
Rothwell (1992)	Interactive innovation	Literature review		
Rothwell and Dodgson (1991)	Networks	Questionnaire	Basic statistics (numbers)	100 UK and 80 Italian SMEs
Senker and Sharp (1997)	Networks and organisation learning	Interviews; 35 + 14	Comparative	Seven paired cases: MNC/ DBF (Europe/US)
Shaw (1998)	Social network theory	Un/semi-structured + participant obs	Mapping + bartering exchange	Five small graphic design firms
Steward and Conway (1998)	Social Networks	Interviews (SS) + documents	Discourse analysis	20 cases — UK & Germany
Swan and Newell (1995)	Innovation diffusion	Semi-structured Interviews and questionnaire	Multiple regression., correlation, descriptive stats	20 interviews + 189 questionnaires
Tidd (1997)	Networks and learning	Literature review		
Tidd (1993)	Technological fusion	Secondary sources	Discursive	7 UK-based MNCs
Teubal *et al.* (1991)	Evolutionary	Secondary sources	Comparative	None

Appendix 1 *(Continued)*

Turpin *et al.* (1996)	Networks and bricoleurs	Secondary sources (not stated)	Structured comparisons	Four paired cases Australia and China
Von Raesfeld Meijer (1998)	Social cognition	Interviews, observation, documents	Socio-cognitive mapping	Case: Den Bosch bus/rail station
Yli-Renko and Autio (1998)	Systemic evolution (embeddedness)	Interviews, firm and secondary data	Discursive — comparative	Five high-tech firms
Zander (1999)	Networks (international)	Patent data [US + Sweden] (1946–1990)	Cluster analysis	24 Swedish MNCs

References

Aldrich, H. and Zimmer, C. (1986) "Entrepreneurship through social networks," in *Art and Science of Entrepreneurship*, D. Sexton and R. Smilor (eds.), Cambridge, Mass., Ballinger.

Allen, T. (1977) *Managing the Flow of Technology: Technology Transfer and the Dissemination of Technological Information within the R&D Organization.* MIT Press, Boston.

Allen, T. (1970) "Communication networks in R&D laboratories," *R&D Management*, **1**(1), 14–21.

Autio, E, (1997) "New technology-based firms in innovation networks symplectic and generative impacts," *Research Policy*, **26**(3), 263–281.

Baptista, R. and Swann, P. (1998) "Do firms in clusters innovate more?" *Research Policy*, **27**(5), 525–540.

Barber, B. (1952) *Science and the Social Order.* Collier Books, New York.

Barley, S. (1990) "The alignment of technology and structure through roles and networks," *Administrative Science Quarterly*, **35**, 61–103.

Barley, S. and Tolbert, P. (1997) "Institutionalization and structuration: Studying the links between action and institution," *Organization Studies*, **18**(1), 93–117.

Belussi, F. and Arcangeli, F. (1998) "A typology of networks: Flexible and evolutionary firms," *Research Policy*, **27**, 415–428.

Birley, S., Cromie, S. and Myers, A. (1991) "Entrepreneurial networks: Their emergence in Ireland and overseas," *International Small Business Journal*, **9**(4), 56–74.

Brandes, U., Kenis, P., Raab, J., Schneider, V. and Wagner, D. (1999) "Explorations into the visualisation of policy networks," *Journal of Theoretical Politics*, **11**(1), 75–106.

Burns, T. and Stalker, G. M. (1961) *The Management of Innovation*. Pergamon, London.

Burt, R. S. (1992) *Structural Holes: The Social Structure of Competitiveness*. Harvard University Press, Cambridge Mass.

Callon, M. (1992) "The dynamic of techno-economic networks," in *Technological Change and Company Strategies: Economic and Sociological Perspectives*, R. Coombs, P. Saviotti and V. Walsh (eds.), Academic Press, London.

Callon, M. (1986) "Some elements of a sociology of translation: Domestication of the scallops of St. Brieuc Bay," in *Power, Action and Belief: A New Sociology of Knowledge?* Sociological Review Monograph 32, J. Law (ed.), Routledge, London.

Camagni, R. (1991) "Introduction: From the local milieu to innovation through cooperation networks," in *Innovation Networks: Spatial Perspectives*, R. Camagni (ed.), Belhaven, London.

Chesbrough, H. W. and Teece, D. J. (1996) "When is virtual virtuous? Organizing for innovation," *Harvard Business Review*, **74**(1), 65–73.

Child, J. and Faulkner, D. (1998) *Strategies of Cooperation: Managing Alliances, Networks and Joint Ventures*. Oxford University Press, Oxford.

Clark, P. and Staunton, N. (1993) *Innovation in Technology and Organization*. Routledge, London.

Clegg, S. and Hardy, C. (1996) "Conclusion: Representations," in *Handbook of Organization Studies*, S. Clegg, C. Hardy and W. Nord (eds.), Sage, London.

Clegg, S., Hardy, C. and Nord, W. (1996) *Handbook of Organization Studies*. Sage, London.

Cohen and Levinthal (1990) "Absoptive capacity: A new perspective on learning and innovation," *Administrative Science Quarterly*, **35**, 128–152.

Conway, S. (1997) "Informal networks of relationships in successful small firm innovation," in *Technology, Innovation and Enterprise: The European Experience*, D. Jones-Evans and M. Klofsten (eds.), MacMillan, London, 236–273.

Conway, S. (1995) "Informal boundary-spanning communication in the innovation process," *Technology Analysis and Strategic Management.* **7**(3), 327–342.

Conway, S. (1994) "Informal boundary-spanning links and networks in successful technological innovation," unpublished Ph.D. dissertation, Aston Business School, Birmingham.

Conway, S. and Shaw, E. (1999) "Networking and the small firm," in *Enterprise and Small Business: Principles, Policy and Practice*, D. Jones-Evans and S. Carter (eds.), Addison Wesley Longman.

Conway, S. and Steward, F. (1998) "Mapping innovation networks," *International Journal of Innovation Management*, **2**(3), 165–196.

Cooke, P. (1996) "The new wave of regional innovation networks: Analysis, characteristics and strategy," *Small Business Economics*, **8**(2), 159–171.

Cooke, P. and Morgan, K. (1994) "The regional innovation system in Baden-Wurtemberg," *International Journal of Technology Management*, **9**(3/4), 394–429.

Coombs, R., Richards, A., Saviotti, P. and Walsh, V. (1996) *Technological Collaboration: The Dynamics of Cooperation in Industrial Innovation.* Edward Elgar, Cheltenham.

Crane, D. (1972) *Invisible Colleges: Diffusion of Knowledge in Scientific Communities.* University of Chicago Press, Chicago.

Czpiel, J. A. (1975) "Patterns of inter-organizational communication and the diffusion of a major technological innovation in a competitive industrial community," *Academy of Management Journal*, **18**(1), 6–24.

DeBresson, C. and Amesse, F. (1991) "Networks of innovators: A review and introduction to the issues," *Research Policy* (special issue), **20**(5), 363–379.

de Laat, P. (1999) "Systemic innovation and the virtues of going virtual: The case of the digital video disk," *Technology Analysis and Strategic Management*, **11**(2), 139–180.

Doz, Y. L., Hamel, G. and Prahalad, C. K.. (1989) "Collaborate with your competitor and win," *Harvard Business Review*, **67**(1), 133–139.

du Gay, P. (1996) *Consumption and Identity at Work*. Sage, London.

Duncan, R. B. (1976) *The Ambidextrous Organization: Designing Structures for Innovation*. North Holland, New York.

Ebers, M. (ed.) (1997) *The Formation of Inter-Organization Networks*. Oxford University Press, Oxford.

Fletcher, D, (1998) "Swimming around in their own ponds: The weakness of strong ties in developing innovative practices," *International Journal of Innovation Management*, **2**(2), 137–160.

Foray, D. (1991) "The secrets of industry are in the air: Industrial cooperation and the organisational dynamics of the innovative firm," *Research Policy*, **20**(5), 393–405.

Foucault, M. (1986) *Power/Knowledge: Selected Interviews and Other Writings 1972–1977*, Colin Gordon (ed.), Harvester, Brighton.

Freeman, C. (1991) "Networks of innovators: A synthesis of research issues," *Research Policy*, **20**(5), 499–514.

Frost, P. and Whitley, R. (1971) "Communication patterns in a research laboratory," *R&D Management*, **1**(2), 71–79.

Garvey, W., Lin, N. and Nelson, C. (1970) "Some comparisons of communication activities in the physical and social sciences," in *Communication Among Scientists and Engineers*, C. Nelson and D. Pollock (eds.), Heath Lexington Books, Massachusetts, 61–84.

Gemser, M, Leenders, A. A. M. and Nachoem, M. W. (1996) "The dynamics of inter-firm networks in the course of the industry life-cycle: The role of appropriability," *Technology Analysis and Strategic Management*, **8**(4), 439–454.

Giddens, A. (1984) *The Constitution of Society: Outline of the Theory of Structuration*. Polity, Cambridge.

Graham, I. (1998) "The construction of a network technology: Electronic livestock auction markets," *International Journal of Innovation Management*, 2(2), 183–200.

Grandori, A. and Soda, G. (1995) "Inter-firm networks: Antecedents, mechanisms, and forms," *Organization Studies*, 16(2), 183–214.

Granovetter, M. (1985) "Economic action and social structure: The problem of embeddedness," *American Journal of Sociology*, 91(3), 481–510.

Granovetter, M. (1973) "The strength of weak ties," *American Journal of Sociology*, 78(6), 1360–1380.

Griffith, B. and Mullins, N. (1972) "Coherent social groups in scientific change," *Science*, 177(4053), 959–964.

Grint, K. and Woolgar, S. (1997) *The Machine at Work: Technology, Work and Organisation*. Polity, Cambridge.

Hagedoorn, J. and Schakenraad, J. (1992) "Leading companies and networks of strategic alliances in information technologies," *Research Policy*, 21(2), 163–190.

Hagedoorn, J. and Schakenraad, J. (1991) "Inter-firm partnerships for generic technologies: The case of new materials," *Technovation*, 11(7), 429–444.

Hakansson, H. (1990) "Technological collaboration in industrial networks," *European Management Journal*, 8(3), 371–379.

Hakansson, H. (1987) "Product development in networks," in *Industrial Technological Development: A Network Approach*, H. Hakansson (ed.), Crook Helm, London.

Herner, S. (1954) "Information gathering habits of workers in pure and applied science," *Industrial and Engineering Chemistry*, 46(1), 228–236.

Hislop, D., Newell, S., Scarbrough, H. and Swan, J. (1997) "Innovation and networks: Linking diffusion and implementation," *International Journal of Innovation Management*, 1(4), 427–448.

Hobday, M. (1994) "The limits of the silicon valley network model: A critique of network theory," *Technology Analysis and Strategic Management*, 6(2), 231–244.

Imai, K. and Baba, Y. (1989) "Systemic innovation and cross-border networks: Transcending markets and hierarchies," *OECD Conference on Science, Technology and Economic Growth*, Paris.

Johannisson, B. and Peterson, R. (1984) "The personal networks of entrepreneurs," *Third Canadian Conference*, International Council for Small Business, Toronto, 23–25 May; citing Ashby, W. (1956) *Introduction to Cybernetics*, Wiley, New York.

Jones, O. (1997a) "Structuration theory and technology transfer: Towards a social theory of innovation," Aston Business School Research Paper, RP9704.

Jones, O. (1997b) "Changing the balance: Taylorism, TQM and work organisation," *New Technology, Work and Employment*, **12**(1), 13–24.

Jones, O., Cardoso, C. C. and Beckinsale, M. (1997) "Mature SMEs and technological innovation: Entrepreneurial networks in the UK and Portugal," *International Journal of Innovation Management*, **1**(3), 201–227.

Jones, O., Conway, S. and Steward, F. (1998) "Introduction: Social interaction and innovation networks," *International Journal of Innovation Management* (special issue), **2**(2), 123–136.

Jones O., Edwards, T. and Beckinsale, M. (2000) "Technology management in a mature firm: Structuration theory and the innovation process," *Technology Analysis and Strategic Management*, **12**(2), 161–177.

Karlsson, C. and Olsson, O. (1998) "Product innovation in small and large enterprises," *Small Business Economics*, **10**(1), 31–46.

Knights, D., Murray, F. and Willmott, H. (1993) "Networking as knowledge work: A study of strategic interorganizational development in the financial services industry," *Journal of Management Studies*, **30**(6), 975–975.

Koschatzky, K. (1998) "Firm innovation and region: The role of space in the innovation process," *International Journal of Innovation Management*, **2**(4), 383–408.

Koski, H. (1999) "The implications of network use, production network externalities and public networking programmes for firms products," *Research Policy*, **28**(4), 423–439.

Kreiner, K. and Schultz, M. (1993) "Informal collaboration in R&D: The formalisation of networks across organisations," *Organisation Studies*, **14**(2), 189–202.

Kuhn, T. (1962) *The Structure of Scientific Revolutions*. University of Chicago Press, Chicago.

Laamanen, T. and Autio, E. (1996) "Global dominant dynamic complementarities and technology-motivated acquisitions of new technology-based firms," *International Journal of Technology Management*, **12**(7/8), 769–786.

Lamming, R. (1993) *Beyond Partnership: Strategies for Innovation and Lean Supply*. Prentice Hall, London.

Langlois, R. N. and Robertson, R. L. (1992) "Networks and innovation in a modular system: Lessons from the microcomputer and stereo components industries," *Research Policy*, **21**(4), 297–314.

Larson, J. and Rogers, E. (1984) *Silicon Valley Fever: Growth of High-Technology Culture*. George Allen and Unwin, London.

Latour, B. (1988) *The Pasteurisation of France*. Harvard University Press, Cambridge.

Latour, B. (1999) "On recalling ANT," in *Actor Network Theory and After*, (eds.) J. Law and J. Hassard, Blackwell, Oxford.

Law, J. (1999) "After ANT: Complexity, naming and topology," in *Actor Network Theory and After*, J. Law and J. Hassard (eds.), Blackwell, Oxford.

Law, J. and J. Hassard (eds.) (1999) *Actor Network Theory and After*, Blackwell, Oxford.

Lawton Smith, H., Dickson, K. and Smith, S. (1991) "There are two sides to every story: Innovation and collaboration within networks of large and small firms," *Research Policy*, **20**, 457–468.

Leonard-Barton, D. (1984) "Interpersonal communication patterns among Swedish and Boston-area entrepeneurs," *Research Policy*, **13**(2), 101–114.

Levitt, B. and March, J. (1988) "Organizational learning," *American Review of Sociology*, **14**, 319–340.

Lin, N., Garvey, W. and Nelson, C. (1970) "A study of the communication structure in science," in *Communication Among Scientists and Engineers*, C. Nelson and D. Pollock (eds.), Heath Lexington Books, Massachusetts, 23–60.

March, J. G. (1991) "Exploration and exploitation in organizational learning," *Organization Science*, **2**, 71–87.

Marquis, D. and Allen, T. (1966) "Communication patterns in applied technology," *American Psychologist*, **21**, 1052–1060.

Martin, J. and Frost, P. (1996) "The organizational culture war games: A struggle for intellectual dominance," in *Handbook of Organization Studies*, S. Clegg, C. Hardy and W. Nord (eds.), Sage, London.

Menzel, H. (1962) "Planned and unplanned scientific communication," in *The Sociology of Science*, B. Barber and W. Hirsch (eds.), Free Press, New York, 417–441.

Midgley, D. F., Morrison, P. D. and Roberts, J. H. (1992) "The effect of network structure in industrial diffusion processes," *Research Policy*, **21**(5), 533–552.

Morgan, G. (1986) *Images of Organization*. Sage, Beverly Hills.

Nelson, R. R. (1990) "Capitalism as an engine of progress," *Research Policy*, **19**, 193–214.

Nelson, R. R. and Winter, S. G. (1982) *An Evolutionary Theory of Economic Change*. Harvard University Press, Cambridge.

Nohria, N. and Eccles, R. G. (eds) (1992) *Networks and Organizations: Structure, Form and Action*. Harvard Business School Press, Boston.

Nonaka, I. and Takeuchi, H. (1995) *The Knowledge Creating Company: How Japanese Companies Create the Dynamics of Innovation*. Oxford University Press, Oxford.

Oliver, A. L. and Ebers, M. (1998) "Networking network studies: An analysis of conceptual configurations in the study of inter-organizational relationships," *Organization Studies*, **19**(4), 549–583.

Piore, M. and Sabel, C. (1984) *The Second Industrial Divide*. Basic Books, New York.

Perry, N. (1993) "Scientific communication, innovation networks and organization structures," *Journal of Management Studies,* **30**(6), 957–973.

Pfeffer, J. and Salancik, G. R. (1978) *The External Control of Organizations: A Resource-Dependence Perspective.* Harper and Row, New York.

Pollard, S. (1965) *The Genesis of Modern Management: A Study of the Industrial Revolution in Great Britain.* Edward Arnold, London.

Porter, M. (1998) *The Competitive Advantage of Nations.* 2nd ed., Macmillan, Basingstoke.

Powell, W. W., Koput, K. W. and Smith-Doerr, L. (1996) "Interorganizational collaboration and the locus of innovation: Networks of learning in bio-technology," *Administrative Science Quarterly,* **41**(1), 116–145.

Powell, W. W. (1990) "Neither market nor hierarchy: Network forms of organization," *Research in Organizational Behavior,* **12**, 295–336.

Price, D. (1963) *Little Science, Big Science.* Columbia University Press, New York.

Price, D. and Beaver, D. (1966) "Collaboration in an invisible college," *American Psychologist,* **21**, 1011–1018.

Roberts, K. and Grabowski, M. (1996) "Organizations, technology and structuring," in *Handbook of Organization Studies,* S. Clegg, C. Hardy and W. Nord (eds.), Sage, London.

Robertson, M., Swan, J. and Newell, S. (1996) "The role of networks in the diffusion of technological innovation," *Journal of Management Studies,* **33**(3), 333–359.

Robertson, P. L. and Langlois, R. N., (1995) "Innovation, networks and vertical integration," *Research Policy,* **24**(4), 543–562.

Rothwell, R. (1992) "Successful innovation: Critical factors for the 1990s," *R&D Management,* **22**(3), 221–240.

Rothwell, R. and Dodgson, M. (1991) "External linkages and innovation In small & medium-sized enterprises," *R&D Management,* **21**(2), 125–137.

Saxenian, A. (1991) "The origins and dynamics of production networks in Silicon Valley," *Research Policy,* **20**(6), 423–438.

Saxenian, A. (1990) "Regional networks and the resurgence of Silicon Valley," *California Management Review*, **33**(1), 89–112.

Schumpeter, J. A. (1943) *Capitalism, Socialism and Democracy*. Harper Row, New York.

Senker, J. and Sharp, M. (1997) "Organisational learning in cooperative alliances: Some case studies in biotechnology," *Technology Analysis and Strategic Management*, **9**(1), 35–52.

Shaw, E. (1998) "Social networks: Their impact on the innovative behaviour of small service firms," *International Journal of Innovation Management*, **2**(2), 201–222.

Sobrero, M. and Schrader, S. (1998) "Structuring inter-firm relationships: A meta-analytic approach," *Organization Studies*, **19**(4), 585–615.

Steward, F. (1995) "Risk analysis and rival technological trajectories: Consumer safety in bread and butter," in *Managing Technology in Society: The Approach of Constructive Technology Assessment*, A. Rip, T. J. Misa and J. Schot (eds.), Pinter, London.

Steward, F. and Conway, S. (1998) "Situating discourse in environmental innovation networks," *Organization*, **5**(4), 479–502.

Steward, F. and Conway, S. (1996) "Informal networks in the origination of successful innovations," in *The Dynamics of Cooperation in Industrial Innovation*, R. Coombs, A. Richards, P. Saviotti and V. Walsh (eds.), Edward Elgar, Cheltenham.

Stinchcombe, A. (1990) *Information and Organization*. University of California Press, Berkeley.

Swan, J. and Newell, S. (1995) "The role of professional associations in technology diffusion," *Organization Studies*, **16**(5), 103–127.

Teubal, M. Yinnon, T. and Zuscovitch, E. (1991) "Networks and market creation," *Research Policy*, **20**(5), 381–392.

Salancik, G. R., (1995) "WANTED: A good network theory of organisation," *Administrative Science Quarterly*, **40**, 345–349.

Thompson, J. D. (1967) *Organisations in Action: Social Science Bases of Administrative Theory*. McGraw-Hill, New York.

Tidd, J. (1997) "Complexity, networks and learning: Integrative themes for research on the management of innovation," *International Journal of Innovation Management*, **1**(1), 1–19.

Tidd, J. (1993) "Technological innovation, organisational linkages and strategic degrees of freedom," *Technology Analysis and Strategic Management*, **5**(3), 273–284.

Turpin, T., Garrett-Jones, S. and Rankin, N. (1996) "Bricoleurs and boundary riders: Managing basic research and innovation knowledge networks," *R&D Management*, **26**(3), 267–282.

Uehara, E. (1990) "Dual exchange theory, social networks, and informal social support," *American Journal of Sociology*, **96**, 523–524.

Van de Ven, A. H., Angle, H. L., and Poole, M. S. (eds.) (1989) *Research on the Management of Innovation*. Harper and Row, New York.

von Raesfeld Meijer (1998) "Missing the bus: A socio-cognitive perspective on technological networks," *International Journal of Innovation Management*, **2**(2), 161–182.

Williamson, O. (1985) *The Economic Institutions of Capitalism*. Free Press, New York.

Williamson, O. (1975) *Markets and Hierachies*. Free Press, New York.

Wolek, F. and Griffith, B. (1974) "Policy and informal communication in applied science and technology," *Science Studies*, **4**, 411–420.

Yli-Renko, H. and Autio, E. (1998) "The network embeddedness of new technology-based firms: Developing a systemic evolution model," *Small Business Economics*, **11**(3), 253–267.

Zald, M. N. (1970) *Power in Organizations*. Vanderbilt University Press, Nashville.

Zander, I. (1999) "How do you mean 'global'? An empirical investigation of innovation networks in the multinational corporation," *Research Policy*, **28**(2/3), 195–213.

Chapter 2

Micropolitics and Network Mapping: Innovation Management in a Mature Firm

Oswald Jones and Martin Beckinsale

Introduction: Innovation and Networks

By the mid-1960s, there was an extensive and coherent US literature concerned with the management of innovation (Allison, 1969;
Bright, 1964; Cole, 1967; Lothrop, 1964; Pelz and Andrews, 1966). In
1967, an R&D Unit was established at the newly-founded Manchester
Business School with a mission to make 'an impact on the theory
and practice of management of research and development' (Pearson,
1988:100). The *R&D Management* journal was founded in 1970 by the
Unit's Director (Alan Pearson) followed in the mid-1970s by an annual
conference focusing on industrial innovation. Furthermore, writers
who have investigated the role of new technology have exerted strong
influence over the development of organisational theory (for example,
Woodward, 1958; Thompson, 1967; Perrow, 1972). Even today, there
are few organisational theory courses that do not make reference
to the influential work of Burns and Stalker (1961). Although, as

41

discussed below, their work is generally misrepresented as belonging to the crude school of contingency theory in which structure is determined by environment (see Hatch, 1997). Since the 1970s, there have been a number of important research projects which have attempted to identify factors which influence successful innovation. These vary from large-scale efforts such as Wealth from Knowledge (Langrish *et al.*, 1972) to case studies describing intrapreneurial success in overcoming corporate indifference (Katz, 1994). Some researchers (SPRU, 1972) have attempted to distinguish between successful and less successful innovators. Studies also scrutinised innovation at the level of the organisation (Clark, 1995), the sector (Walsh, 1984) as well as the so-called national systems of innovation (Lundvall, 1992; Nelson, 1993; Porter, 1990; Edquist, 1997). As a result of this substantial body of research, the *interactive model of innovation* (Rothwell and Zegveld, 1985) has replaced traditional linear models such as 'science-push' (Schumpeter, 1934; 1943) or 'market-pull' (Schmookler, 1966).

In recent years, it has been increasingly acknowledged in the literature that networks are central to effective innovation management. While it is recognised that innovation networks have existed for many years: 'Information technology has led to the widespread diffusion of modes of networking which were previously far less common' (Freeman, 1991:510). Similarly, Rothwell's (1992) 'fifth generation model' of innovation management emphasises the importance of IT as a facilitator of inter-organisational linkages. Hence, the need to develop a revised research agenda: 'With growing complexity, a focus on the role of innovation networks will be more appropriate than the behaviour of specific firms in isolation' (Tidd, 1997:16). At the same time, Granovetter (1985) points out, all activity is embedded in complex networks of social relations which include family, state, educational and professional background, religion, gender and ethnicity. Certainly, we believe that it is impossible to consider decision-making without examining networks of personal relationships. Managers do not simply examine 'transaction costs' when making decisions about the purchase of materials, components or technologies (Williamson, 1975; 1985). As pointed out by Sako (1992), *trust* between individual employees and a firm's customers and suppliers is an essential feature of business relationships. Combining the institutional links associated with transaction cost analysis and the social processes identified by actor-network theorists means that 'there is some potential for convergence in the future development of networks and alliances' (Coombs *et al.*, 1996:8).

In our view, the major failing of the network approach is that there is usually no attempt to analyse the role played by internal politics. The exception are scholars such as Callon (1986) and Latour (1988) who developed the concept of actor-network theory. Politics are central to the dynamic act of 'convergence' when networks are created: 'Sometimes there is controversy, conflict, and the translation is rejected as betrayal: *traduttore-traditore*. We find workers who do not want to play the role defined for them by the machine; consumers who doubt the quality and value of a product; scientists who denounce the arguments of fellow authors...' (Callon, 1991:144). Other writers who discuss links between networks and political action generally focus on the way in which inter-organisational power is exercised as a means of accessing or securing resources (Keil *et al.*, 1997; Eig and Johansson, 1997). Krackhardt and Hanson (1993) examine the way in which political actions influence innovation but they see the mapping of informal networks as a managerial tool. As discussed in Jones *et al.* (1998), a number of research projects at Aston Business School have been at the forefront of establishing the validity of *mapping* as a legitimate approach to the study of innovation (Steward and Conway, 1998; 1996; Conway, 1994; 1995). However, in our view this work also fails to acknowledge the political and power dimensions which are associated with innovation. Krackhardt and Hanson (1993:104) posit that the shift from formal hierarchies to flatter, organic structures means that informal networks are analogous to an organisation's central nervous system. Mapping these informal networks is essential if we are to 'uncover the source of political conflicts and failure to achieve strategic objectives' (Krackhardt and Hanson, 1993:106).

Simply identifying the political actors may not provide sufficient information to anticipate the overall dynamics of political influences on innovation (Krackhardt, 1992). Network mapping provides a method for precisely identifying political actors and the way in which they influence decisions. The approach is also useful for revealing ways in which organisational structure influences access to power. Over recent years, there has been increasing standardisation in the techniques associated with mapping innovation networks. Conway (1997) provides a network template which helps identify the information flows and transaction content (goods and artifacts) which link various actors. Network templates also encourage a longitudinal approach because it is possible to accurately compare changing structures and network arrangements. Hence, the political actors and political decision makers can be mapped and the outcomes of their political decisions evaluated. We believe that the micropolitical actions associated with acquiring resources for successful innovation are not susceptible managerial intervention:

> "Under these conditions, organizations can be viewed as loose structures of interests and demands, competing for organizational attention and resources, and resulting in conflicts that are never completely resolved" (Narayanan and Fahey, 1982:26).

In this chapter, we focus on the way in which micropolitics influenced the development of informal and formal networks in a 'mid-corporate' manufacturing company over a six-year period. We define micropolitics as the way in which managers pursue their own individual career interests by attempting to gain political influence and status (Burns, 1961; Narayanan and Fahey, 1982; Knights and Murray, 1994). Although as pointed out by Thomas (1994) engineers and scientists involved in the innovation process can also be powerful political actors. Few authors, even those concerned with the politics of innovation (see Fincham *et al.*, 1994), discuss what Burns, introducing the third edition, describes as his 'preoccupation with the structure and dynamics of interpersonal relations' (Burns and Stalker, 1994:xiii). In fact, Burns goes on to say that the distinction between organic and mechanistic structures owed at least as much to 'internal politics' as it did to changes in the external environment (also see Burns, 1961:267).

The Politics of Innovation

Given the internal conflict which usually accompanies innovation, it is surprising that the political dimension has received little attention from those concerned with innovation management. Although Burns and Stalker (1961) are widely quoted, the sociological sophistication of the 'mechanistic-organic' polarity which places managerial ambition at the centre of innovation management is largely ignored. As Webb (1992:473) points out, two variables other than the rate of environmental change affect structure: the relative strength of individual commitments to gaining political influence and status; secondly, the capacity of senior managers to interpret the external environment and take appropriate action. Recently, Thomas (1994) made an important attempt to incorporate political processes into his model of innovation. Mainstream R&D writers are also criticised by Cabral-Cardoso (1996:47) because of claims that project selection: 'should be conducted according to well-defined rules and procedures normally translated into mathematical models or other formalised techniques' (see Souder, 1972). According to Cabral-Cardoso (1996:55), project selection cannot be fully understood without considering internal politics:

"Managers are seen as political actors who use technical information as a means of influencing decision outcomes with a view to satisfying their own self-interests."

The political implications of technological innovation are most in evidence in literature concerned with the labour process (Braverman, 1974). Much of this work is overtly politicised as it builds on Marxist conceptions of class exploitation (Thompson, 1989). Perhaps surprisingly, the more mainstream organisational behaviour (OB) literature also tends to ignore political processes. Although the recently published *Handbook of Organization Studies* (Clegg, *et al.*, 1996) contains contributions by many writers regarded as critical organisational theorists (Alvesson, Burrell, Calas, Clegg, Reed, Smircich and Townley) the 730 pages contain very few references to 'politics'. 'Power' is treated more seriously *via* a review of the literature from traditional Marxist and Weberian perspectives to more recent views associated with gender, identity, resistance, knowledge and emancipation (Hardy and Clegg, 1996). While Miller *et al.* (1996) examine the links between power and decision-making in organisations. The authors point out that issues associated with who is involved in the making of decisions 'are central to the politics of organizational behaviour' (Miller *et al.*, 1996:294).

Others argue that power and politics are located in the deep structure of organisations which influence the interpretative schemes and cognitive maps of individuals and groups (Frost and Egri, 1990a/b). Political influence on innovation is demonstrated by re-examining a number of well-known cases including post-it-notes, the Dvorak keyboard and Tagamet. It is concluded that political activity is most likely to take place if the proposed innovation threatens the self-interest of 'a powerful coalition'.

"In the case of product innovations, the hallmark of failure appears to be the intense political activity at both the surface and deep structure levels. Interests opposing an innovation marshalled a wide range of tactics to preserve the status quo" (Frost and Egri, 1990b:6).

In our view, attempting to identify the political influences on innovatory activity from secondary sources is problematic. According to Frost and Egri (1990b:7), both the post-it-note and Tagamet involved 'relatively little political conflict'. At the same time, the authors acknowledge that there was internal opposition to both projects as well as considerable bargaining for resources.

Kleinschimdt and Cooper (1995) posit 'managerial perceptions' as the key in explaining many innovation failures. Questionnaire data were collected on 103 major new product innovations (successes and failures) in the chemical industry over a two-year period. Managers in the survey underrated the importance of activities such as business analysis, initial screening, field trials and market launch activities and overrated technical factors such as product development, in-house testing and technical assessment. In addition, for every 100 days spent on industrial projects, 78 were spent on technology-related issues compared to only 12 devoted to marketing (mostly concentrated on product launch). Although the study provides an important insight into organisational influences on innovation, suggestions that technical activities such as product development and laboratory testing in the chemical industry are 'relatively low' in their impact on financial success are unconvincing. This is because the authors rely entirely on correlation coefficients to establish relationships between variable values and the degree of project success. As Kleinschimdt and Cooper (1995:287) point out 'only association and not causality can be inferred' as well as noting that 'multivariate analysis was also undertaken, but the results *proved more difficult to present*' (op cit, 298) [emphasis added].

In a related area, Piercy (1986:255) criticises the traditional marketing literature because there is little explicit recognition of power and politics. The study uses an information-structure-power model (Pfeffer, 1981) to explain the role of political power in relation to resource allocation. The empirical data were based on a questionnaire survey of 'marketing directors' in 600 UK manufacturing firms which provided 166 useable returns (29% response rate). Three 'indicators' were used to assess political behaviour: other departments access to marketing information; marketings success in accessing information held by other departments; and, the extent to which marketing shared responsibility for sales forecasting (Piercy, 1986:302). Each indicator was operationalised by a series of questions and associated Likert scales. Although Piercy (1986:467) accepts his work has limitations, he is still positive about the study's value:

> "While these measures do not predict all variance in resources (at best the models show R^2 of only 0.50) it is suggested that these findings are of some significance to the management of companies…and hence of some note for management teachers and researchers."

We believe it is worth repeating a question posed by Gareth Morgan (1983:12): 'What lies on the other side of the correlation coefficient?' It is not reliability of Piercy's findings with which we take issue but the extent to which quantitative data are capable of revealing the subtly of managers pursuing their own interests whether related to the setting of budgets or with innovation. The basis of our argument is that micropolitics do have an extremely powerful influence on innovatory activity. Such activity can only be revealed by employing longitudinal, qualitative studies of organisational activity. Methods based on the re-interpretation of existing cases, questionnaire surveys or even interviews are unlikely to reveal the full range of political activity occurring in organisations.

Micropolitics and Innovation Networks

Organisational politics (OP) are based on the strategies that managers and employees pursue to secure their careers, avoid blame, create success and establish stable identities (Knights and Murray, 1994). Such a definition combines elements of agency and structure

(Giddens, 1984): agents are concerned with improving their career prospects as well as their physical and mental well-being. However, all such activity takes place within the broader societal structures including class, gender, social institutions and organisations. Knights and Murray (1994) suggest that Pettigrew's (1973) work was a 'major achievement' because sociological conceptions of power and politics were used to analyse organisational decision-making. Pettigrew utilised two concepts to develop his analytical framework, first, the idea of organisational sub-units competing for scarce resources (Cyert and March, 1963). Secondly, the politics of career advancement as a key element influencing organisational activities (Mills, 1956; Dalton, 1959; Burns and Stalker, 1961). Knights and Murray (1994) argue that there has been very little subsequent work which examines organisations from a political perspective (Noble, 1977; Wilkinson, 1983; Scarbrough and Corbett, 1992; Thomas, 1994). The political perspective was applied to the implementation of information systems (IS) by Pfeffer (1981) and Mintzberg (1983). It was found that various actors pursued political strategies designed to control the organisation's ideological and structural resources which often resulted in visible conflict. Although this literature did question rationalist views of organisational life, it was weakened by 'an assumption that an absence of observable conflict means an absence of politics' (Knights and Murray, 1994:15).

According to Knights and Murray (1994:28), there are two main problems with the existing literature related to politics and organisational change. First, the liberal pluralist tradition of perceiving politics as a deviant, irrational activity which disrupts the smooth running of organisations. Hence, there is a tendency to ignore the broader structural issues associated with social inequality. Secondly, politics are regarded as an 'uncomplicated reflection' of class and gender issues arising from the capitalist political economy. Consequently, many observers tend to dismiss micropolitical manoeuvring as an aberration which occurs in the absence of effective managerial control. An alternative view is that political activity takes place within the apparently cooperative set of managerial practices associated with everyday coordination and control:

> "Disruptive political behaviour is a consequence of previous exercises of power that have failed to secure the appropriate support from others" (Knights and Murray, 1994:30).

Tidd *et al.* (1997:306) acknowledge there are a range of influences on innovation and network relationships: visionary leadership (Kay, 1995), structure (Burns and Stalker, 1961), key individuals (Rubenstein, 1994), teamworking (Wheelwright and Clark, 1992), extensive communications (Allen, 1977), high involvement (Bessant and Caffyn, 1996), customer focus (Rothwell, 1992), innovatory climate (Jones, 1996), learning organisation (Leonard-Barton, 1995). We believe that the micropolitics associated with resource allocation are the key influence on innovatory activity in most organisations. The majority of managers are acutely aware that career competition, conflict and the formation of alliances are central to their everyday activities but they still emphasise the rationality of their own activity. Some believe that political actions are divergent from formal goals and tend to be associated with open conflict (Drory and Romm, 1990). This view is contested by Knights and Murray (1994) who suggest that organisational politics (OP) are central to organisational activities and that individual political objectives are often concealed by the pursuit of legitimate organisational goals:

> "The process of OP, and the vast array of individual and collective strategies it involves, is necessarily riven through with a central paradox. For a great deal of managerial practice constructs a reality of its own activity that denies the political quality of that practice" (Knights and Murray, 1994:31).

Politics associated with individuals attempting to further their own careers (or inhibit the careers or others) are seen as the norm, rather than an abnormality, of organisational life (see Mills, 1956; Dalton, 1959). According to Burns and Stalker (1994:xvii) the micropolitics of managerial ambition are intrinsically linked to innovation and organisational change:

> "Careerism and internal politics were visibly present in every one of the sixteen or so firms, English and Scottish, that the study eventually covered. Nor did they seem necessarily to resolve themselves either in orderly settlement or in reconciling differences by reconstructing the firm's policy."

Such a view is supported by Thomas (1994:17) who, in developing his 'political perspective', points out organisations are social constructs which 'can be amended, dissolved, and reconstructed through the actions of their participants'. At the same time, the 'routines' which are embedded in organisational structures help provide a predictable context in which innovation takes place (Nelson and Winter, 1982) .

Bloor and Dawson (1994) draw on Giddens (1984) to explain the way in which professional and organisational cultures interact (Fig. 1). This framework is helpful for illustrating the way in which innovation is influenced by internal politics, organisational structure and culture, as well as factors in the external environment. Individual political activity occurs as managers pursue their career interests in competition with rivals and, occasionally, in conflict with stated organisational objectives. Equally, various functions and departments within the organisation (marketing, R&D, finance, production/service delivery) attempt to exert influence over the allocation of resources for the innovation of new products.

All micropolitical activity occurs within a broader historical context which includes the founder's vision, values and beliefs as well as

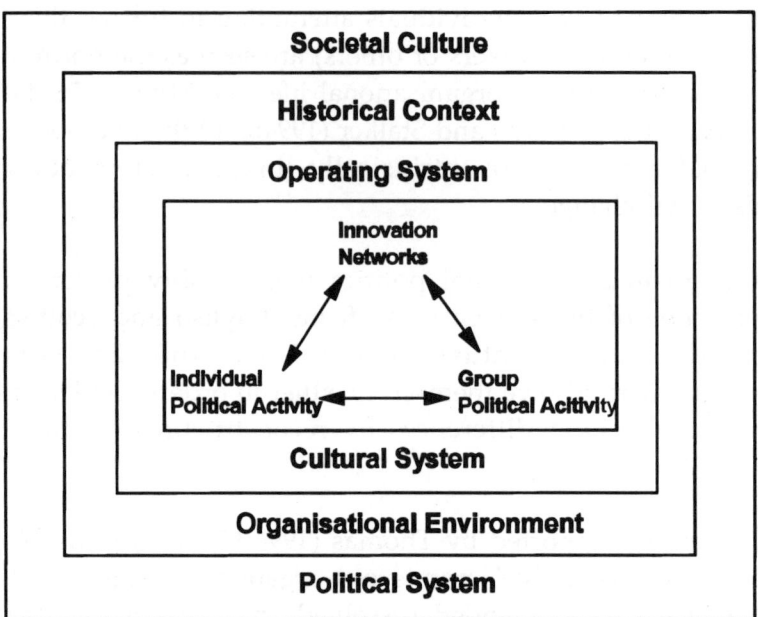

Fig. 1. Micropolitics and innovation networks.

previous markets, structures and technologies. Organisational environment refers to suppliers, customers and competitors operating in the market sectors in which business is conducted. The operating system includes structures, technologies and personnel systems adopted by the organisation while the cultural system incorporates the configuration of beliefs, values, ceremonies and stories which are shared by organisational members. Societal culture is represented by the norms, beliefs, values, contemporary life-styles and social expectations of the population at large (Bloor and Dawson, 1994:279). An important addition is the national political system because government policies influence organisational activity by creating an institutional framework which has been described as a 'national system of innovation' (Lundvall, 1992; Nelson, 1993; Porter, 1990; Edquist, 1997). While we acknowledge these broader influences, our argument is that micropolitics are central to the innovation process rather than an aberration which occurs because of poor communications or inadequate managerial control.

Research Methods

This research was initiated in 1992 by a group project which fulfilled the coursework requirements of an MBA elective on technology strategy. Funding was then acquired from the DTI to study PAC's approach to innovation management which was identified as 'best practice' (Jones and Beckinsale, 1994). Subsequently, regular interviews have been carried out with all the key actors including the technical director and other senior managers and engineers working on major innovation projects. We believe that our approach overcomes one of the major weaknesses associated with most research into innovation and networks. As Freeman (1991:511) states when setting out a research agenda:

"Longitudinal case studies would enable us to gain a better understanding of such complex issues as power relations within networks. Some authors have stressed the rather equal relationships within networks, whilst others have pointed to the tendency for the stronger firms to exploit their position at the core of the networks."

Given that the innovation process takes time, methodologies which do not include a temporal dimension are unlikely to provide a thorough account of organisational activity. Longitudinal research remains rare in organisational studies and single cases do raise issues of generalisability. In discussing the shift from micro to macro levels, Hamel *et al.* (1993:35) argue that it is the objective of the study that is important rather than the number of confirmatory cases. This refers to the distinction between statistical generalisation, in which inference is made about a specific population, and analytical generalisation, in which empirical data are compared with a theoretical 'template' (Yin, 1994:30). Therefore, we believe that the relationship between micro-political actions and organisational change can only be revealed by utilising detailed qualitative methodologies (see Yin, 1993). Examining the way in which political actions influenced innovation networks in PAC meant that it was necessary to establish trust between researchers and our contacts in the company. Over a long period of time, we have been able to observe and record organisational changes and then discuss our interpretations of events with key actors. Equally, because of the strong personal relationships we are able to ask individual managers and engineers to explain their own motivation for certain activities.

PAC the privately-owned company which is the subject of this longitudinal study was founded in 1946, is now located on five UK sites and in 1998 had a annual turnover of approximately £30 million with a work force of 750 employees. The business is based on the manufacture of control devices for the automotive and small domestic appliance industries. PAC is a successful exporter and supplies products to over 40 countries including leading Japanese and German auto manufacturers. DC (Direct Current) cutouts for the automotive industry account for well over 50% of PACs turnover and represents 14% of the world market which is contested with two main rivals. Kettle controls account for most of the remaining turnover and this is also a highly concentrated market in which PAC has only one direct competitor. A further 10% of PAC's turnover is derived from AC (Alternating Current) cutouts which are used in a wide range of small domestic appliances such as battery chargers, cable reels, hair dryers, transformers and washing machine motors. For over 50 years, the manufacture of products utilising thermostatic bi-metallic devices has been the basis of PACs competitive success. The principle of bi-metallic cutouts is used in a wide range of industries but there has been a strong commitment to in-house product development (incremental innovation) to ensure that PAC's products remain competitive. In addition, product development, processes development, and production were closely integrated which encouraged a rapid response to customer requests and was central to PAC's ability to compete with much larger rivals. In the early 1990s, managers recognised that the emphasis on incremental innovation meant that the company was dependent on an ageing technology. Although market share was expanding the major motor manufacturers were able to use high levels of competition to drive down prices and in 1992 a programme was instigated to access new skills and technologies.

Innovation Networks in PAC

At the time the research began in 1992, the firm had a traditional functional structure in which the technical director (Bill Black) was responsible for engineering, electronics, production and quality. At this time, there was a major change in strategy as senior managers sought to end their reliance on bi-metallic technology. Black was given responsibility for identifying new technologies and he set up collaborative projects with two universities in the north west of England. An *ad hoc* group of engineers emerged who had an interest in developing the new product (Appendix 1). The first project followed a proposal from an independent inventor who had patented a bi-metallic device for sensing the pressure of pneumatic tyres. The commercial potential seemed obvious but PAC engineers felt that the technology was too primitive to be acceptable to leading automobile manufacturers: 'A bi-metallic device spinning in a tyre is unstable and provides poor sensor readings even at speeds as low as 30 mph' (PAC engineer). University A was commissioned to develop a more sophisticated tyre monitoring system (TMS) utilising transmitter/receiver technologies (Jones and Smith, 1995). Gradually, it became clear that the TMS would be based on electronic technologies rather than bi-metallic

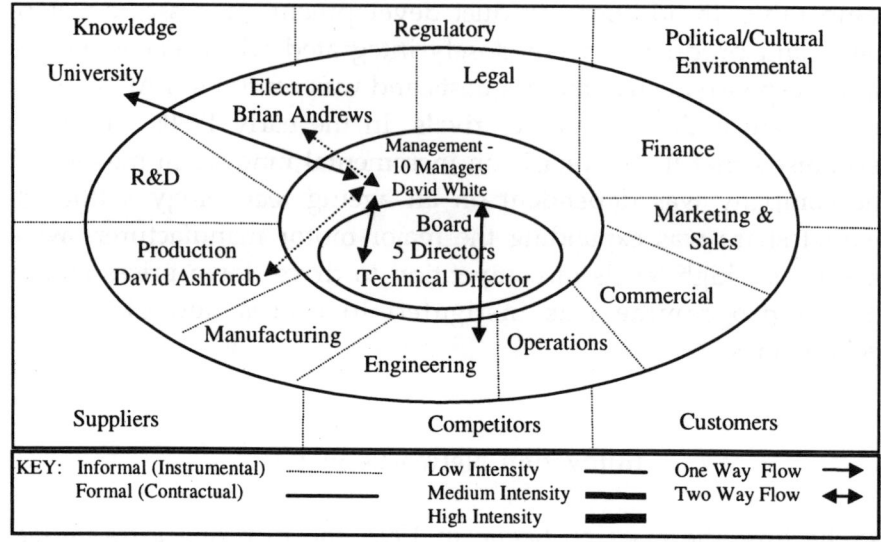

Fig. 2. First network.

strips. Consequently, a network was established that linked the technical director, mechanical and electrical engineers and the University (Fig. 2). The second alliance involved a depth of anaesthesia monitor (DAM) developed by University B to ensure that patients were not conscious during surgery. The University originally commissioned PAC as a manufacturing partner after obtaining project support through the European-ESPRIT programme. PAC eventually took over the design work because the University's device was unreliable and difficult to manufacture.

The map in Fig. 2 is based on the work of Conway (1994; 1997) and illustrates the network structure at the very early stages of the TMS project. The links were still largely informal because there had been no clear decision to proceed with the project. However, the Board had established a formal agreement with the University to evaluate the TMS's feasibility. The network shows that at this stage David White was the project's focal actor and he provided progress reports for the Board. Initially, internal cohesion within the network was low and most development work was carried out by the University.

It proved far more complex than expected to develop the transmitter-receiver technology central to the TMS. PAC engineers felt that they had been misled about the level of expertise within the University: 'They claimed they had state-of-the-art knowledge when in fact they were starting from scratch'. Bill Black decided to end links with the university and concentrate resources by incorporating the TMS project into the electronics department (Appendix 2). Mechanical engineers continued to work on general engineering projects as well as contributing towards development of the TMS. At this stage, the functional structure meant that individual project managers were responsible for domestic appliance and automotive products. For example, David White was managing the TMS project as well as a number of other motor products. By early 1995, the TMS project team included: David Ashford (product development engineer), Brian Andrews (assistant electronics manager), Ken Charles (development engineer) and Guy Marks (product manager). Figure 3 illustrates the network when the TMS project was incorporated into the electronics department. Individual actors were still not full-time and retained links to their original departments which meant the TMS network remained 'open'. In addition, there was no concerted effort to identify and develop a market for the product. At this stage, David White remained the project's focal point and was linked to the Board through Bill Black. Gradually, the number of internal and external actors increased and multiple role relations became more complex. All communication and information flows were two-way except for the ESPRIT programme which set milestones for the TMS team.

Senior managers felt there was a need to improve communications between marketing, engineering and production: 'The engineering departments have become vast and sprawling with the result that engineers have lost touch with our customers' (Bill Black). In

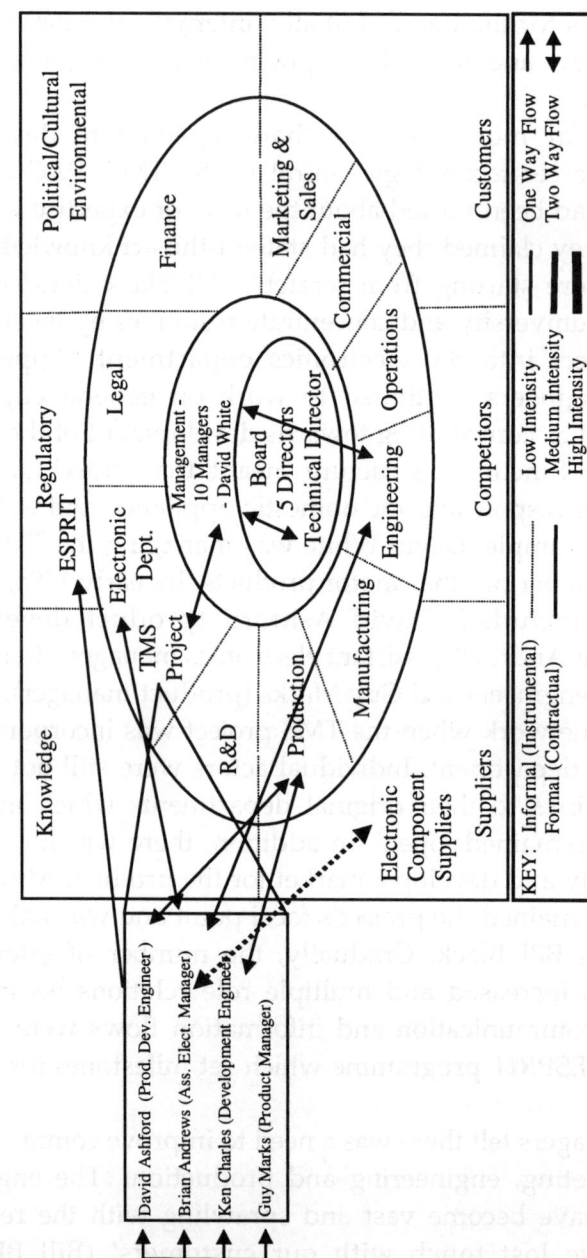

Fig. 3. Second network.

mid-1995, the functional structure based on engineering, accounts, sales and production was reorganised into three divisions: automotive products, domestic appliances and operations. At the same time, four product managers were appointed in sales to create a clearer customer focus. Two engineers, reporting directly to the technical director, were given managerial responsibility for automotive products and domestic appliances. The newly-appointed automotive manager, David White, was in charge of motor products and electronic products which included the TMS. Airflow products, water heating and sensor products were directly supervised by the domestic appliance manager (Appendix 3). The objectives were to give electronics a stronger strategic focus and to consolidate PAC's position in existing markets by making more effective use of in-house skills to develop new products.

Restructuring meant that engineers worked on single projects within either the automotive or domestic appliance groups which encouraged new sets of relationships based on individual products (Fig. 4). Links with external actors remained unchanged but the *ad hoc* matrix arrangement, in which engineers divided their loyalties between function and project, was replaced by product-based networks. In particular, those working on the TMS project concentrated on strong linkages within the team rather than maintaining relationships with those in production and engineering. The TMS team were no longer diverted by the competing requirements of other product development projects:

"Yes it's better now we're all full-time on this project...and are directly responsible to David White" (Brian Andrews).

Employees from a range of functions, production, engineering and development, were located in each of the five product areas. The creation of product managers in sales also encouraged the integration of knowledge and expertise across the organisation. Despite these

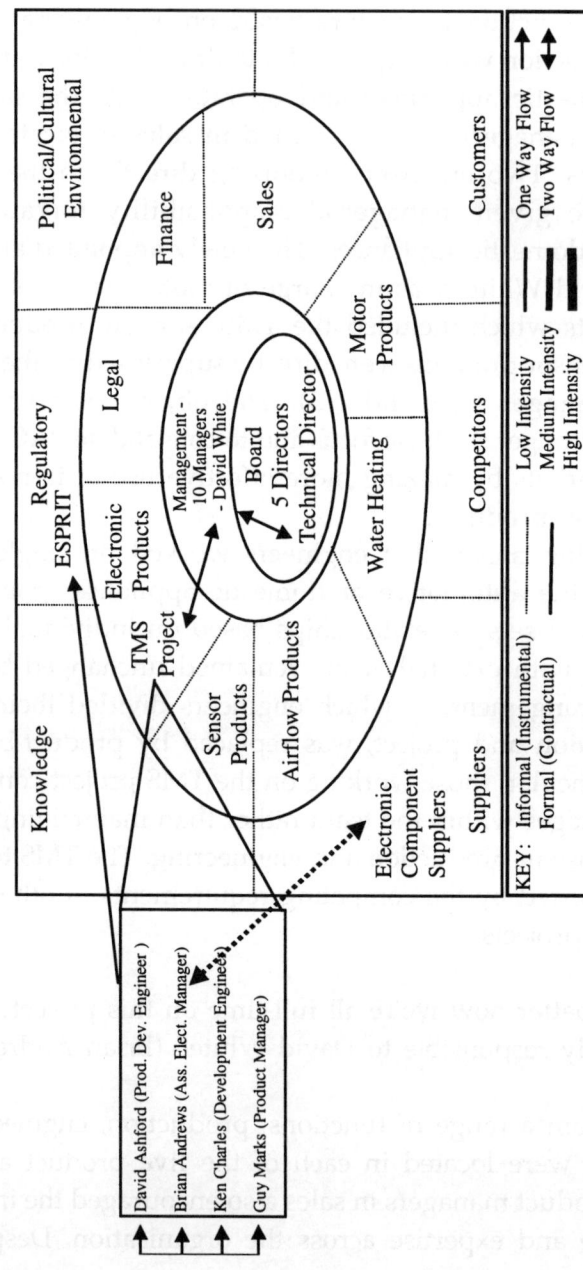

Fig. 4. Third network.

changes, it became apparent late in 1996 that the TMS team were having difficulty in developing reliable battery technologies and there was no progress in convincing leading auto manufacturers to adopt the product as standard equipment. As a consequence, Bill Black, under pressure from the board, decided that the project should be terminated. For contractual reasons related to the ESPRIT Programme it was necessary to retain one engineer on the project for a further six months.

In January 1997, PAC underwent another restructuring as senior managers decided to return to a functional form of organisation. Bill Black was appointed managing director and his existing role was split between domestic appliances and industrial controls. After six months, David White was promoted to the position of technical director (Appendix 4). This change was a reaction to the failure to produce an effective prototype of the DAM and because of difficulties developing the TMS into a marketable product. The technical director had previously been responsible for tool-making which was closely integrated with product design helping ensure that new tools and manufacturing equipment were developed simultaneously with new products. Shifting responsibility for tool-making to the operations director meant that there was a clear distinction between product and process innovation. Another significant change was that the electronics department was disbanded and 15 electronic engineers were made redundant.

The Micropolitics of Innovation Networks

Over the last nine years, there have been three major structural changes which have re-organised relationships within PAC (Appendix 1–4). The original functional structure gave way to a product-based organisation and the recently revised structure incorporates aspects of both product and function. The new arrangement means that there is now networking across functional as well as product boundaries. Mapping the changes in relationships provides a deeper understanding of the way in which the innovation process actually works. Our major criticism of those utilising the network approach is that they do not acknowledge that micropolitics influence social interaction. In the following section, we argue that individuals pursuing their own career objectives were the main factor in the changing networks within PAC.

There were two key political acts which created informal networks within the organisation and which eventually led to structural change. First, the technical director's decision to develop two extremely radical products: the TMS and the DAM. There was considerable resistance within the organisation not least from the financial director who was anxious about issues of 'customer liability' associated with the new products. This led to a 'healthy debate' between Bill Black, who believed that new opportunities must be seized and the financial director who stated that 'the DAM in particular is too far removed from our existing markets'. Bill Black managed to overcome the doubts of other directors by pointing to the massive potential for PAC if either of the two products were successfully developed. Subsequently, an informal network emerged which consisted of employees who strongly believed in the value of the TMS. As other senior managers within the organisation gradually recognised that product could be successfully brought to market the network was formalised in an electronic products department.

The second political act concerned David White's commitment to the TMS after its cancellation. By late 1996, optimism had disappeared as a series of technical problems accentuated the difficulty of successfully bringing the TMS to market. Bill Black was promoted to the position of managing director and his role was split between domestic appliances and industrial controls to give greater market focus. Following the project's termination, White retained one development engineer for contractual reasons:

"We had to continue with the development of the TMS in order to meet the requirements of the ESPRIT Programme."

Despite the closing of the electronics departments, David White also managed to divert two electronic engineers to work *informally* on the TMS. This arrangement, which he described as 'a skunkworks' (Quinn, 1985) continued for a further nine months during which time another project team made a breakthrough in battery-life development. Furthermore, two other factors encouraged his belief that the TMS could legitimately be revived. First, a leading Formula 1 Team expressed interest in the product. Secondly, the company's insurers eventually decided that the problems associated with 'customer liability' could be overcome without PAC incurring punitive premiums. White's decision to keep faith with the TMS was vindicated and late in 1997 his 'skunkworks' was integrated into the organisational structure (Appendix 4).

The new TMS team included engineers, electronic experts, a design expert as well as marketing and sales expertise which were central to developing a product which would be attractive to the marketplace. David White retained responsibility for the project but acted as a co-ordinator rather than manager. At the same time, there was still considerable doubt amongst project engineers about the extent of top management commitment:

"When the TMS was resurrected a project plan was drawn up with specific targets and dates. Simply, the Board have kept out of the running of the project and will only act if the final completion date for the product is not met — then they will cancel the TMS!" (Ken Graham).

Along with the structural change there were also major changes in the personnel responsible for the TMS and only Ken Charles (Testing) remained on the team. Marks and Andrews were part of the group that were made redundant and Ashford was transferred to another

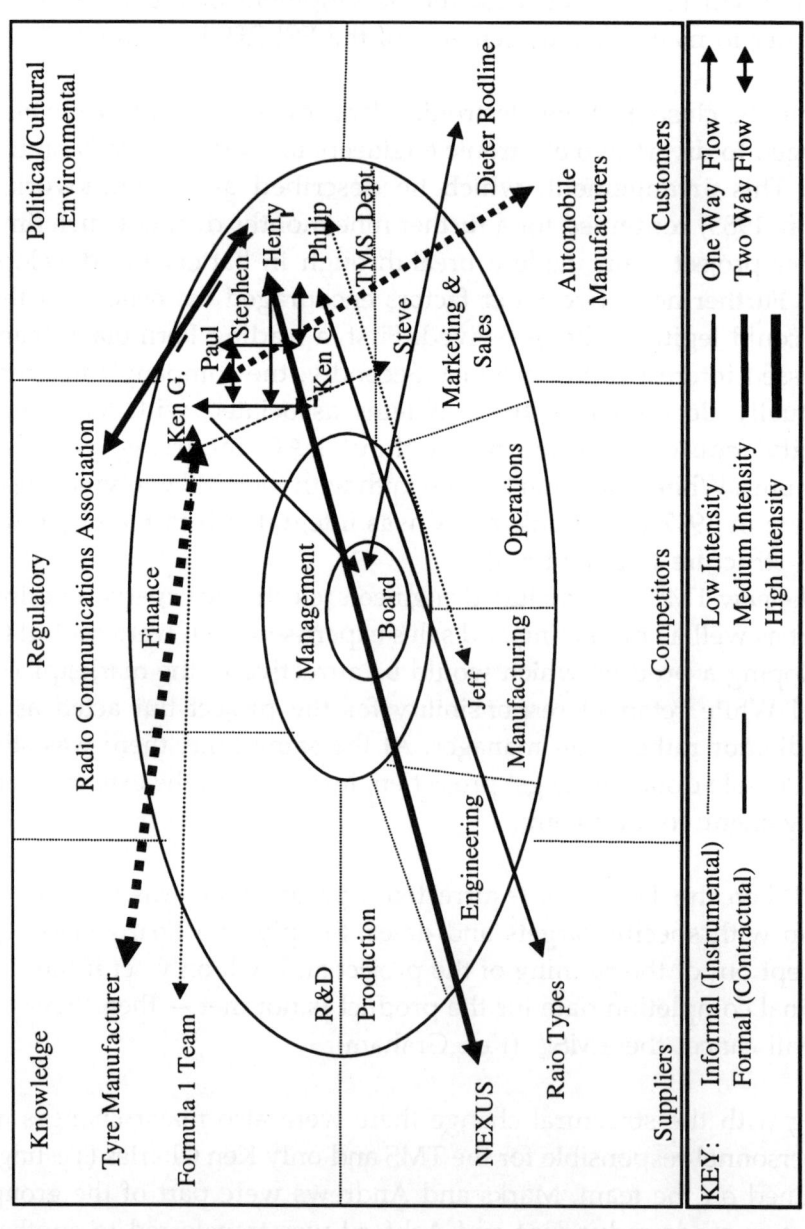

Fig. 5. Fourth network.

department. Engineers with a broader range of skills were recruited to the team in an attempt to ensure that the project was successful. Ken Graham, who joined PAC in 1995, was an electronics engineer with expertise in software development and he was appointed project leader. Paul Martin, a sales engineer, became project coordinator because of his strong links with the automotive sector. Philip Edwards (manufacture/design), a CAD designer working on automated machines, was drafted onto the team to develop the TMS housing. Stephen Jeremy (electronics/software); and Henry Michael (electronics) were transferred from electronic products to the TMS team.

As Fig. 5 illustrates, the TMS network underwent considerably change as new actors brought in skills appropriate to the revised project requirements. The new network highlights the increasing importance of developing a marketing strategy *via* the links with Steve Brown the sales and marketing manager. This is also illustrated by the strong informal links with a tyre manufacturer and automotive manufacturers. The TMS department became autonomous and White no longer operated as the focal actor. Multiple internal role relations decreased as actors concentrated on the TMS which reduced inter-departmental links. However, external multiple role relations increased with many informal links being developed by sales and marketing.

This new department brought together engineers with the appropriate knowledge to fully develop the TMS. Stephen Jeremy and Henry Michael both had skills in the area of battery technology as well as software development which were key to the TMS. The inclusion of Paul Martin in the new department was particularly significant because he had close ties with tyre manufacturers and large US motor manufacturers. Therefore, he brought a much sharper marketing focus to the project which was in contrast to the technology-driven approach associated with the previous network arrangements.

We do not suggest that either of these political actions occurred simply to benefit the careers of Bill Black and David White. As Knights and Murray (1994) point out, individual political actions need not be at odds with overall organisational goals. However, the original decision by Bill Black was taken in the face of considerable opposition from other senior managers. Also, White's action was highly politicised as he was aware that the project had been officially terminated. Our argument is that studying the processual elements of innovation networks helps reveal that political activities are central to the formation of networks and to decisions about which particular technological trajectories will be pursued within a company. In response to a question about the potential damage to his career by working on the discontinued TMS, David White responded:

> "At the time the risk of proceeding seemed small because I still believed that the TMS was a viable product. Also, in PAC career progression is planned 20 years in advance and promotion does not rely on the success or failure of individual projects."

As illustrated in Fig. 1, micropolitical activity occurs within a broader historical context which includes the founder's vision, values and beliefs as well as market structures and technologies. The historical context of PAC continues to exert leverage over both operating and cultural systems. The owners have a strong sense of social responsibility and have created employment by deliberately locating their factories in rural areas. At the same time, private ownership means that the company is not subject to direct pressures from the stock market. For example, PAC sustained considerable financial losses in the first three years of the 1990s without resorting to job-cuts or reduced investment in new technologies. Broader issues associated with societal culture and the national political system also influenced employees and managers in PAC. The election of New Labour in June 1997 caused considerable alarm because of Blair's commitment to join the European Social Chapter. This was seen as a threat to PAC which relies on some degree of labour market flexibility. Another broader political issue which had implications for the company was the strength of the pound. Over 85% of bimetallic cutouts are exported 'with 25% of sales going to the two most competitive and quality conscious markets in the world, Japan and Germany' (David White). This has made business extremely difficult for PAC and the possibility of manufacturing in the Far East has been under serious consideration. While we acknowledge that these wider societal and political factors are important to companies such as PAC the key issue in this chapter is the way in which micropolitics influences networking arrangements.

Conclusions

In this paper, we argue that much of the literature associated with network mapping fails to acknowledge the political dimension. As Van Maanen (1979) stressed a map is not the territory and this is certainly the case as far as innovation networks are concerned:

> "So, when we map we miss. We miss the gap between representation and image represented. We miss the contrivance of the representational practices that produce the effect of representation. We miss the point if we think that what we see is what we see" (Clegg and Hardy, 1996:676).

What we have tried to do in presenting the case of PAC is to suggest that while mapping networks can be illustrative they provide an extremely partial representation of organisational reality. In examining the changes over a six-year period it can be seen that there is a close relationship between the informal structure represented by networks and an organisation's formal structure (see Burns, 1977:308). We see networks as subjective structures that are inseparable from their social context and the activities of social actors. In fact, even the formal structure of organisations cannot be observed as objective reality. As pointed out by Giddens (1984) social systems do not have structures but they exhibit structural properties. The relationship between individual actions (agency) and organisational structure is similar to the links between grammar and speech. The rules of grammar are utilised by social actors in their patterns of communication which in turn constantly recreates the structure of language. We believe that it is wrong to assume that networks can be studied as objective social structures which are independent of human agency (see Roberts and Grabowski, 1996).

In the early 1990s, senior managers in PAC implemented an ambitious strategy to acquire and utilise radical new technologies. We have mapped the organisational changes which occurred over a six-year period as a result of efforts to develop a tyre pressure sensor. Utilising the work of those such as Burns and Stalker (1961) and Knights and Murray (1994) who have developed a micropolitical perspective on organisational change we have analysed the influences on the emergence of innovation networks in PAC. Two key actors, Bill Black (technical director) and David White exerted considerable political influence within the company during this longitudinal study. Bill Black had to overcome the resistance of fellow directors to ensure that he was able to devote resources to the DAM and the TMS. While White refused to accept that cancellation of the TMS meant that the project was definitely finished. He arranged that two engineers should spend part of their time unofficially working on the project. Fortunately for both actors, early in 1999 it appeared that the TMS at least would make it to the marketplace. Certainly the careers of Bill Black, who is now managing director and David White, who is now technical director, have prospered since 1992. We are not suggesting that either believed that their commitment to these projects would improve their careers. However, what we do believe is that innovation cannot be studied in isolation from the micropolitics which typify all organisations.

Appendix 1: 1992 Structure

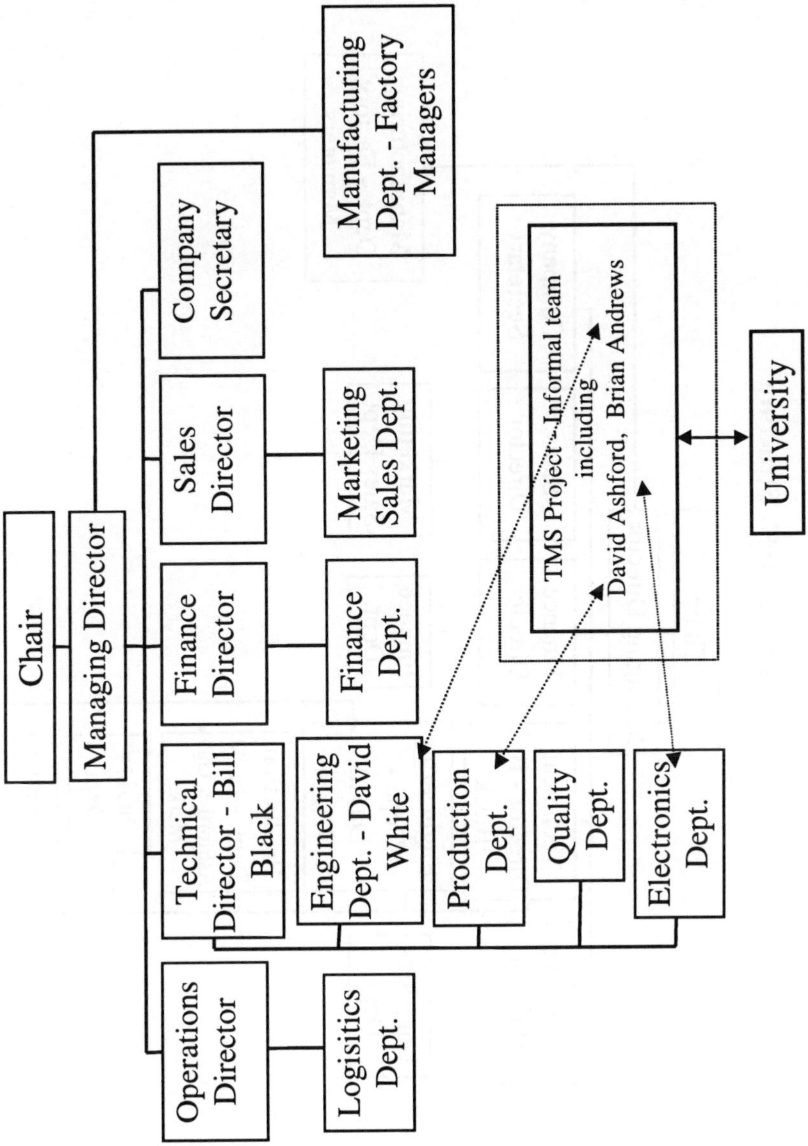

Chair

Managing Director

Operations Director

Technical Director - Bill Black

Finance Director

Sales Director

Company Secretary

Manufacturing Dept. - Factory Managers

Logisitics Dept.

Engineering Dept. - David White

Production Dept.

Quality Dept.

Electronics Dept.

Finance Dept.

Marketing Sales Dept.

TMS Project - Informal team including David Ashford, Brian Andrews

University

Appendix 2: 1993 Structure

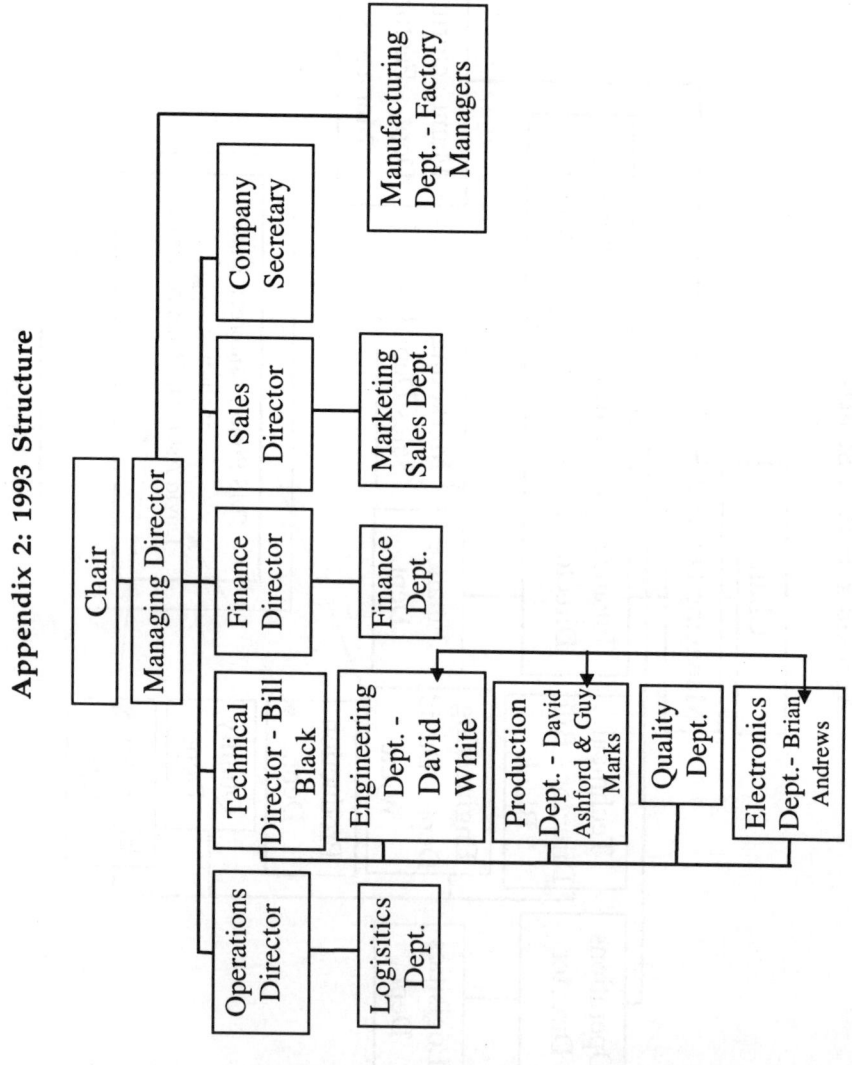

Appendix 3: Mid-1995 Structure

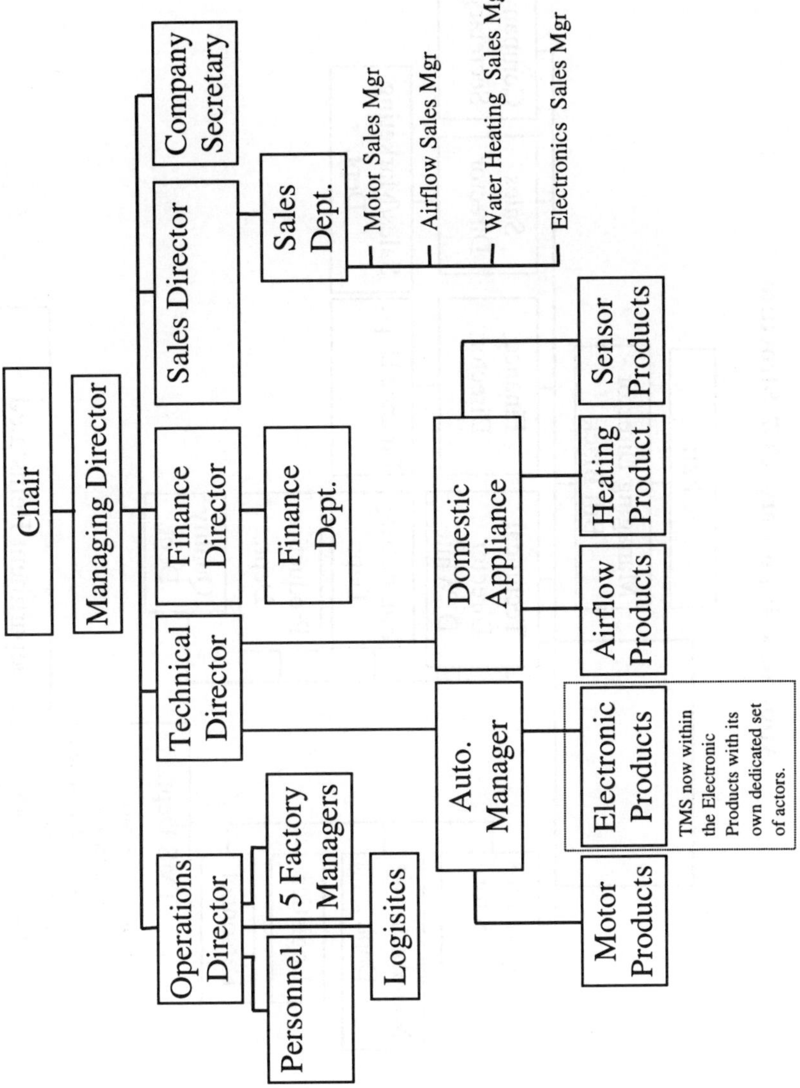

Chair

Managing Director

Operations Director — Technical Director — Finance Director — Sales Director — Company Secretary

Personnel | 5 Factory Managers

Logisitcs

Finance Dept.

Sales Dept.

Motor Sales Mgr

Airflow Sales Mgr

Water Heating Sales Mgr

Electronics Sales Mgr

Auto. Manager — Domestic Appliance

Motor Products | Electronic Products | Airflow Products | Heating Product | Sensor Products

TMS now within the Electronic Products with its own dedicated set of actors.

Appendix 4: January 1997 Structure

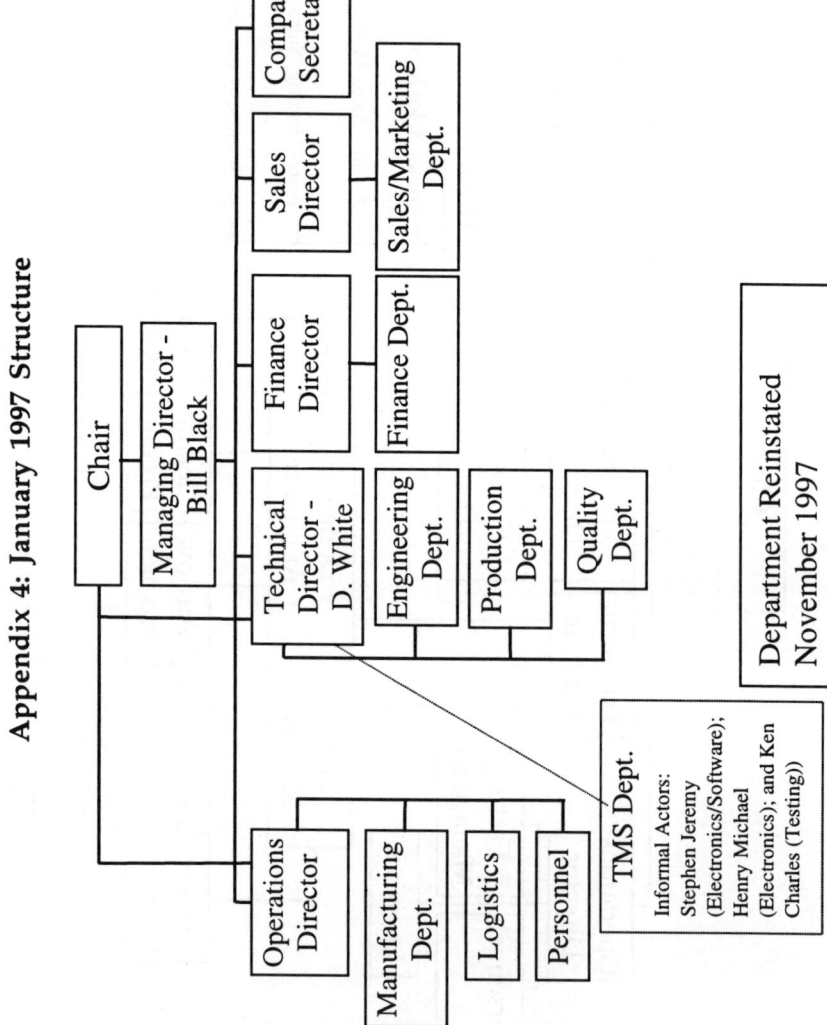

Chair

Managing Director - Bill Black

Company Secretary

Sales Director — Sales/Marketing Dept.

Finance Director — Finance Dept.

Technical Director - D. White
- Engineering Dept.
- Production Dept.
- Quality Dept.

Operations Director
- Manufacturing Dept.
- Logistics
- Personnel

TMS Dept.

Informal Actors:
Stephen Jeremy (Electronics/Software); Henry Michael (Electronics); and Ken Charles (Testing))

Department Reinstated November 1997

References

Allen, T. (1977) *Managing the Flow of Technology: Technology Transfer and the Dissemination of Technological Information within the R&D Organization.* MIT Press, Boston.

Allison, D. (ed.) (1969) *The R&D Game.* MIT Press, Massachusettes.

Bessant, J. and Caffyn, S. (1996) "High involvement innovation," *International Journal of Technology Management,* **14**(1), 121–134.

Bloor, G. and Dawson, P. (1994) "Understanding professional culture in organizational context," *Organization Studies,* **15**(2), 275–295.

Braverman, H. (1974) *Labour and Monopoly Capital.* Monthly Review Press, New York.

Bright, J. R. (1964) *Research, Development and Technological Innovation.* Homewood, Ill.

Burns, T. (1977) *The BBC: Public Institutions and the Private World.* Macmillan, London.

Burns, T. (1961) "Micropolitics: Mechanisms of institutional change," *Administrative Science Quarterly,* **6**, 257–281.

Burns, T. and Stalker, G. M. (1994) *The Management of Innovation.* 3rd ed., Oxford University Press, Oxford.

Burns, T. and Stalker, G. M. (1961) *The Management of Innovation.* Pergamon, London.

Cabral-Cardoso, C. (1996) "The politics of technology management: Influence and tactics in project selection," *Technology Analysis and Strategic Management,* 8(1), 47–58.

Callon, M. (1991) "Techno-economic networks and irreversibility," in *A Sociology of Monsters: Essays on Power, Technology and Domination,* J. Law (ed.), Routledge, London.

Callon, M. (1986) "Some elements of a sociology of translation: Domestication of the scallops of St. Brieuc Bay," in *Power, Action and Belief: A New Sociology of Knowledge?* Sociological Review Monograph 32, J. Law (ed.), Routledge, London.

Clark, J. (1995) *Managing Innovation and Change: People, Technology and Strategy.* Sage, London.

Clegg, S. and Hardy, C. (1996) "Conclusion: Representations," in *Handbook of Organization Studies*, S. Clegg, C. Hardy and W. Nord (eds.), Sage, London.

Clegg, S., Hardy, C. and Nord, W. (1996) *Handbook of Organization Studies.* Sage, London.

Cole, R. I. (ed.) (1967) *Improving Effectiveness in R&D.* London, Academic Press.

Conway, S. (1997) "Focal innovation action sets: A methodological approach for mapping innovation networks," *Aston Business School Research Paper,* RP9702.

Conway, S. (1995) "Informal boundary-spanning communication in the innovation process," *Technology Analysis and Strategic Management,* 7(3), 327–342.

Conway, S. (1994) *Informal Boundary-Spanning Links and Networks in Successful Technological Innovation.* Unpublished Ph.D. Dissertation, Aston Business School, Birmingham.

Conway, S. and Steward, F. (1998) "Mapping innovation networks," *International Journal of Innovation Management,* 2(3), 165–196.

Coombs, R., Richards, A, Saviotti, P. and Walsh, V. (1996) *Technological Collaboration: The Dynamics of Cooperation in Industrial Innovation.* Edward Elgar, Cheltenham.

Cyert, R. M. and March, J. G. (1963) *A Behavioural Theory of the Firm.* Prentice-Hall, Englewood Cliffs.

Dalton, M. (1959) *Men Who Manage.* Wiley, New York.

Drory, A. and Romm, T. (1990) "The definition of organisational politics: A review," *Human Relations,* 43(11), 1133–1154.

Edquist, C. (ed.) (1997) *Systems of Innovation.* Pinter, London.

Eig, U. and Johansson, U. (1997) "Decision making in inter-firm networks as a political process," *Organization Studies,* 18(3), 316–384.

Fincham, R., Fleck, J., Procter, R., Tierney, M., Williams, R. and Scarborough, H. (1994) *Expertise and Innovation: Information Technology in the Financial Services Sector.* Oxford University Press, Oxford.

Freeman, C. (1991) "Networks of innovators: A synthesis of research issues," *Research Policy,* **20**(5), 499–514.

Frost, P. J. and Egri, C. P. (1990a) "Influence of political action on innovation: Part I," *Leadership and Organizational Development Journal,* **11**(1), 17–25.

Frost, P. J. and Egri, C. P. (1990b) "Influence of political action on innovation: Part II," *Leadership and Organizational Development Journal,* **11**(2), 4–11.

Giddens, A. (1984) *The Constitution of Society.* Cambridge: Polity.

Granovetter, M. (1985) "Economic action and social structure: The problem of embeddedness," *American Journal of Sociology,* **91**(3), 481–510.

Hamel, J., Dufour, S. and Fortin, D. (1993) *Case Study Methods* (Qualitative Research Methods Series 32), Sage, London.

Hardy, C. and Clegg, S. (1996) "Some dare call it power,"in *Handbook of Organization Studies,* S. Clegg, C. Hardy and W. Nord (eds.), Sage, London.

Hatch, M. J. (1997) *Organization Theory: Modern, Symbolic and Postmodern Perspectives.* Oxford University Press, Oxford.

Jones, O. (1996) "Human resources, scientists and internal reputation: The role of climate and job satisfaction," *Human Relations,* **49**(3), 269–294.

Jones, O. and Beckinsale, M. (1994) "Alliances between SMEs and HEIs: Technology management in otter controls," *Report for the DTI Innovation Unit.* DTI, London.

Jones, O., Conway, S. and Steward, F. (1998) "Introduction: Social interaction and innovation networks," *International Journal of Innovation Management* (special issue), **2**(2), 123–136.

Jones, O. and Smith, D. (1995) "Strategic technology management in a mid-corporate firm: The case of PAC," *Aston Business School Research Paper,* RP9511.

Katz, R. (1994) "Managing high performance R&D teams," *European Management Journal,* **12**(3), 243–252.

Kay, J. (1995) *The Foundations of Corporate Success: How Business Strategies Add Value*. Oxford University Press, Oxford.

Keil, T., Autio, E. and Robertson, P. (1997) "Embeddedness, power, control and innovation in the telecommunications sector," *Technology Analysis and Strategic Management*, **9**(3), 299–316.

Kleinschmidt, E. J. and Cooper, R. G. (1995) "The relative importance of new product success determinants — Perception versus reality," *R&D Management*, **25**(3), 281–298.

Knights, D. and Murray, F. (1994) *Managers Divided: Organizational Politics and Information Technology Management*. Wiley, Chichester.

Krackhardt, D. (1992) "The strength of strong ties: The importance of philos in organizations," in *Networks and Organizations*, N. Nohria and R. G. Eccles (eds.), Harvard Business School, Boston.

Krackhardt, D. and Hanson, J. R. (1993) "Informal networks: The company behind the chart," *Harvard Business Review*. July/August, 104–111.

Langrish, J., Gibbons, M., Evans, W. and Jevons, F. R. (1972) *Wealth From Knowledge*. Macmillan, London.

Latour, B. (1988) *The Pasteurisation of France*. Harvard University Press, Cambridge.

Leonard-Barton, D. (1995) *Wellsprings of Knowledge: Building and Sustaining the Sources of Innovation*. Harvard Business School Press, Boston.

Lothrop, W. C. (1964) *Management Uses of R&D*. Harper & Row, New York.

Lundvall, B.-A. (ed.) (1992) *National Systems of Innovation*. Pinter, London.

Miller, S. J., Hickson, D. J. and Wilson, D. C. (1996) "Decision-making in organisations," in *Handbook of Organization Studies*, S. Clegg, C. Hardy and W. Nord (eds.), Sage, London.

Mills, C. W. (1956) *White Collar Work*. Oxford University Press, New York.

Mintzberg, H. (1983) *Power in and Around Organizations*. Prentice-Hall, Englewood Cliffs.

Morgan, G. (ed.) (1983) *Beyond Method*. Sage, Beverly Hills.

Narayanan, V. K. and Fahey, L. (1982) "The micro-politics of strategy formulation," *Academy of Management Review*, 7(1), 25–34.

Nelson, R. (1993) *National Systems of Innovation*. Blackwell, Oxford.

Nelson, R. and Winter, S. (1982) *An Evolutionary Theory of Economic Change*. Harvard University Press, Boston.

Noble, D. F. (1977) *America by Design: Science, Technology and the Rise of Corporate Capitalism*. Alfred A. Knopf, New York.

Pearson, A. (1988) "Twenty-one years of research into the management of R&D," *R&D Management*, 19(2), 99–101.

Pelz, D. C. and Andrews, F. M. (1966) *Scientists in Organisations: Productive Climates for R&D*. Wiley, New York.

Perrow, C. (1972) *Complex Organisations: A Critical Essay*. Scott Freeman, New York.

Pettigrew, A. (1973) *The Politics of Organizational Decision-Making*. Tavistock, London.

Pfeffer, J. (1981) *Power in Organizations*. Pitman, Boston.

Piercy, N. (1986) *Marketing Budgeting: A Political and Organisational Model*. Croom Helm, London.

Porter, M. (1990) *The Competitive Advantage of Nations*. Free Press, New York.

Quinn, J. B. (1985) "Managing innovation: Controlled Chaos," *Harvard Business Review*, 73(3), 73–84.

Roberts, K. and Grabowski, M. (1996) "Organizations, technology and structuring," in *Handbook of Organization Studies*, S. Clegg, C. Hardy and W. Nord (eds.), Sage, London.

Rothwell, R. (1992) "Successful industrial innovation: Critical success factors for the 1990s," *R&D Management*, 22(3), 221–239.

Rothwell, R and Zegveld, W. (1985) *Reindustrialisation and Technology*. Longman, New York.

Rubenstein, A. (1994) "Ideation and entrepreneurship," in *Managing New Technology Development*, W. Souder and J. Sheriman (eds.), McGraw-Hill, New York.

Sako, M. (1992) *Prices, Quality and Trust: Inter-Firm Relationships in Britain and Japan*. University Press, Cambridge.

Scarbrough, H. and Corbett, M. (1992) *Technology and Organisations: Power, Meaning and Design*. Routledge, London.

Schmookler, J. (1966) *Invention and Economic Growth*. Harvard University Press, Boston.

Schumpeter, J. A. (1943) *Capitalism, Socialism and Democracy*. Harper Row, New York.

Schumpeter, J. A. (1934) *The Theory of Economic Development*. Harvard University Press, Cambridge, Mass.

Science Policy Research Unit (1972) *Success and Failure in Industrial Innovation*. Centre for the Study of Industrial Innovation, London.

Souder, W. E. (1972) "A scoring methodology for assessing the suitability of management models," *Management Science*, 18:B526-42.

Steward, F. and Conway, S. (1998) "Situating discourse in environmental innovation networks," *Organization*, 5(4), 483–506.

Steward, F. and Conway, S. (1996) "Informal networks in the origination of successful innovations," in *The Dynamics of Cooperation in Industrial Innovation*, R. Coombs, A. Richards, P. Saviotti and V. Walsh (eds.), Edward Elgar, Cheltenham.

Thomas, R. J. (1994) *What Machines Can't Do: Politics and Technology in the Industrial Enterprise*. University of California Press, Berkeley.

Thompson, J. D. (1967) *Organisations in Action: Social Science Bases of Administrative Theory*. McGraw-Hill, New York.

Thompson, P. (1989) *The Nature of Work: An Introduction to Debates on the Labour Process*. 2nd ed., Macmillan, London.

Tidd, J. (1997) "Complexity, networks and learning: Integrative themes for research on the management of Innovation?" *International Journal of Innovation Management*, 1(1), 1–19.

Tidd, J, Bessant, J. and Pavitt, K. (1997) *Managing Innovation: Integrating Technological, Market and Organisational Change*. Wiley, Chicester.

Van Maanen, (1979) "Reclaiming qualitative methods for organisation research: A preface," *Administrative Science Quarterly*, **24**(4), 520–526.

Walsh, V. (1984) "Invention and innovation in the chemical industry: Demand pull or discovery push?" *Research Policy*, **13**, 211–234.

Webb, J. (1992) "The mis-management of innovation," *Sociology*, **26**(3), 471–492.

Wheelwright, S. and Clark, K. (1992) *Revolutionising Product Development*. Free Press, New York.

Wilkinson, B. (1983) *The Shopfloor Politics of New Technology*. Heinemann, London.

Williamson, O. (1985) *The Economic Institutions of Capitalism*. Free Press, New York.

Williamson, O. (1975) *Markets and Hierarchies*. Free Press, New York.

Woodward, J. (1958) *Management and Technology*. HMSO, London.

Yin, R. K. (1994) *Case Study Research: Design and Methods*. 2nd ed., Thousand Oaks, Sage.

Yin, R. K. (1993) *Applications of Case Study Research*. Sage, Newbury Park.

Chapter 3

Employing Social Network Mapping to Reveal Tensions Between Informal and Formal Organisation

Steve Conway*

Introduction

The recognition of the importance of informal or social networks within organisations can be traced at least as far back as the late 1930s (Barnard, 1938; Roethlisberger and Dickson, 1939). Indeed, Blau and Scott (1962:6) argue that "it is impossible to understand the nature of formal organization without investigating the networks of informal relations". More recently, Krackhardt and Hanson (1993:104) have argued:

"Many executives invest considerable resources in restructuring their companies, drawing and redrawing organizational charts only to be disappointed by the results. That's because much

*E-mail: s.conway@aston.ac.uk

of the real work of companies happens despite the formal organization...informal networks can cut through formal reporting procedures to jump start stalled initiatives and meet extraordinary deadlines."

Some researchers go further, positing that "work organisations are...basically sets of social relations for doing work" (Whitley, 1977:56). In recent years, the innovation studies literature has almost universally espoused the virtues of informal organisation (Allen, 1977; Kanter, 1985; Freeman, 1991; Kreiner and Schultz, 1993; Shaw, 1993; Conway, 1995). Informal networks are seen as an important device for promoting communication, integration, flexibility, and novelty, within and between organisations. They are viewed as structures that supplement, complement and add value to the formal organisation. A smaller number of these authors have also highlighted the potential for tension or conflict between the activities of the informal and formal organisation. This tension is largely seen to manifest itself in the potential for dissonance in the goals and objectives of these two 'structures' (Rogers, 1982; Mansfield, 1985; Hippel, 1987; Carter, 1989). The implication is that the utility of the informal organisation can be harnessed if it is 'brought into line' with the goals and objectives of the formal organisation (Allen, 1977; Schrader, 1991). This perspective assumes that the informal organisation is essentially a misdirected but benign creature.

In contrast, the organisation studies literature has long provided a more critical perspective; this recognises the complex interplay and interweaving of the formal and informal organisation (Burns and Stalker, 1961; Blau and Scott, 1962; Blauner, 1964; Gouldner, 1964; Mouzelis, 1967; Tichy *et al.*, 1979; Pfeffer, 1981; Burns and Flam, 1987). Others have highlighted the dysfunctional dimension of informal networks (Tichy, 1981; Shrum, 1990; Krackhardt and Hanson, 1993; Stacey, 1993; Burns and Stalker, 1994). Krackhardt (1993:104), for example, argues that "informal networks can just as easily sabotage companies' best laid plans by blocking communication and fomenting opposition to change", while Stacey (1996:341), contends that:

> "The informal organisation is essentially destabilising. It exists, sometimes in place of the formal organisation and sometimes in competition with it...the informal...organisation is the way in which the formal organisation is changed and it is so changed in the most fundamental way by altering the existing paradigm."

However, Tichy (1981:227) notes that "despite pervasive references to emergent structures in organizations...there has been very little systematic theorising. Even fewer empirical studies of emergent structures exist". This view is supported by Freeman (1991:500–502), who argues that "although rarely measured systematically...informal networks are extremely important, but very hard to classify and measure".

Drawing upon social network analysis, recent research at Aston is attempting to re-emphasise and re-locate both the 'micro-social' (Conway, 1995; 1997a; Steward and Conway, 1996) and 'micro-political' (Jones and Beckinsale, 1999; Jones and Stevens, 1999) activity within organisations back at the very heart of our understanding of the innovation process. For example, Jones and Stevens (1999:175) argue that:

> "The various 'sectional interests' of groups and individuals becomes particularly apparent during the NPD [new product development] process. Reputations, and consequently career prospects, can be enhanced or ruined according to the success or failure of a new product or service. Therefore, any attempt to set out a framework for NPD must include explicit recognition of the role played by micro-politics."

The objective of this chapter is to develop a deeper understanding of 'informal' or 'social organisation', and to outline a methodological approach to help reveal informal networks. In this sense, it is presented as a conceptual rather than an empirical paper, although a network case-study is employed to both illustrate the method and the phenomenon. The chapter is comprised of four main sections: the first provides an overview of the social network literature and sets out an approach for studying informal organisation, termed 'network mapping'; the second highlights the key distinctions between formal and informal organisation and attempts to account for the emergence of the latter; the third explores the tensions between the formal and informal organisation, such as dissonance in objectives and goals, and their existence as competing power-bases; the fourth presents a case-study of the inter-play between the formal and informal organisation in the implementation of an IT system. The formal and informal relationships mobilised at various stages of the implementation process are mapped out to highlight the dynamic nature and role of these linkages. While the case-study reaffirms the importance of informal networks, in this instance in overcoming technical problems in the implementation of an IT system, it also highlights the tension between the formal and informal organisation. Although the formal organisation stood to benefit greatly from the mobilisation of the informal network, these interactions were perceived as undermining the legitimate power and personal credibility of key actors in the formal organisation. As a consequence, the informal activity was marginalised, and to a degree 'criminalised', as the formal organisation sought to reassert itself.

It is worth noting that within this chapter (but also more generally in the literature) the terms 'prescribed' or 'formal' network are used synonymously with the term 'formal organisation' (i.e., the organisational chart). The terms 'emergent', 'informal', or 'social' network, as well as 'informal' or 'social' organisation are also used synonymously. However, in developing a more fine-grained notion of informal organisation, a distinction is made between informal organisation (i.e., task related relationships) and social organisation (i.e., non-task related relationships).

Social Network Analysis and Network Mapping

Relational Data and the Social Network Perspective

Relational data are central to the principal concerns of the sociological tradition, where the emphasis is upon the investigation of the 'structure of social systems' (Marsden and Lin, 1982; Scott, 1991). However, there exists a range of views of what constitutes a 'social system'. According to Boulding (1985), the broadest possible definition of a system is "anything that is not chaos". Lundvall (1992) is more specific, viewing a system as constituted by a number of elements and the relationships between those elements. Katz and Kahn (1966) and Weick (1969) argue that organisations too, can be viewed as social systems with relatively stable patterns of interaction. However, Tichy *et al.* (1979:507) contend that:

> "Such a model of organising, if it is to move beyond the metaphorical stage, requires a coherent framework and accompanying methods of analysis that are capable of capturing both prescribed and emergent processes."

Social network analysis provides such a coherent framework, being concerned with the structure and patterning of relationships and seeking to identify both their causes and consequences. Indeed, Fombrun (1982:280–281) argues that "network analysis is a powerful means of describing and analysing sets of units by focusing explicitly on their inter-relationships", which are "seen as *embedded* in a context that both constrains and liberates". In addition, Tichy *et al.* (1979:510) contend that the framework provided by the social network approach "makes it possible to compare and contrast formally prescribed and emergent relationships in a coherent fashion".

The origins of the network perspective may be traced to the structural concerns of the British anthropologist Radcliffe-Brown and the German 'gestalt' tradition of social psychology, principally associated with Kohler. Since this period, in the 1920s and 1930s, a diversity of strands have intersected, fused and once again diverged at various times. From this complex history, Scott (1991) identifies three main lines: the sociometric analysts (typified by Moreno 1934; 1953), who produced a number of technical advances by adopting the methods of graph theory; the Harvard researchers of the 1930s (including Mayo), who explored patterns of interpersonal relations and the formation of 'cliques'; and the Manchester social anthropologists of the 1950s and 1960s, in particular, Barnes, Bott and Mitchell. A diverse range of disciplines continue to contribute to the conceptual development of the network perspective, including: sociology, anthropology, political science and organisation theory (Tichy *et al.*, 1979).

Rogers (1987) distinguishes between two main research traditions of network research: 'relational' and 'structural'. Relational analysis evolved out of the Moreno-type network sociometry of the 1930s to 1950s. While scholarly interest in structural analysis was sparked by the development of 'block-modelling' techniques by White and others at Harvard University in the mid-1970s (Boorman and White, 1976; White *et al.*, 1976). Relational network analysis essentially focuses on the pathways in networks and entails identifying the cliques of individuals among the members of a network. In contrast, structural network analysis focuses on patterns of similarity in relational configurations and entails identifying 'blocks' of actors. Two actors are said to be 'structurally equivalent', and thus in the same block, if they have the same (or similar) pattern of relationships with other members of the system who occupy the same position (Boorman and White, 1976; White *et al.*, 1976; Burt, 1980).

Rogers (1987:11) argues that in the relational research tradition "structure emerges from communication among a set of individuals". This view is supported by Monge and Eisenberg (1987) who argue that in the relational approach, structure grows out of persistent patterns of communication rather than structure prescribing how individuals should communicate. However, while Alba (1982:63) argues that "neither approach excludes the other.... Indeed, they ultimately complement each other", Blau (1982) suggests that the relational approach is more useful in studying recently-formed systems and the structural approach more useful in studying well-established systems. This paper adopts the relational approach to studying social networks.

Mathematical Versus Metaphorical Orientations: Locating the Middle Ground

The 'network' metaphor is a powerful way of viewing organisations. It changes the imagery from a focus on pairs of dyadic relationships to one of "constellations, wheels, and systems of relationships" (Auster, 1990:65) and of 'webs' of group affiliations (Simmel, 1955). Since the 1950s, a number of specialists have sought to devise more formal translations of the network metaphor. By the early 1970s, an avalanche of technical work had lead to the development of a range of highly mathematical social network techniques. Scott (1991) argues that it is from this research that the key concepts of social network analysis have emerged. However, Rogers (1987:14) warns that "far too much, I fear, we admire mathematical elegance in our network tools and tool-makers, while largely ignoring what useful objects we can dig up with these tools". Indeed, Wellman (1983:156) argues that network analysis should be viewed "as a broad intellectual approach, and not as a narrow set of methods".

Research on networks in social science have on the whole emphasised either a descriptive or metaphorical orientation addressing qualitative aspects, or a mathematical orientation drawing upon quantitative statistical methods. Indeed, there has been a polarisation between these two approaches. Shrum and Mullins (1988) cite Collins (1974), whose work they see as 'imagery without technique' and Breiger (1976), whose approach they argue is 'primarily a methodological exercise', as representative of these two extremes. Research into the process of technological innovation at Aston

Business School has sought to steer a path between these two extremes, whilst drawing and building upon the ideas and concepts of these dominant approaches (Conway, 1994; Conway and Steward, 1998a; Steward and Conway, 1998; Jones and Beckinsale, 1999). Thus, in contrast to much of the innovation network research undertaken in recent years, the interest has been concerned as much with notions of 'network as method' as with 'network as a phenomenon'. As a consequence, work has focused on developing creative ways of researching and mapping innovation networks (Conway, 1997b; Conway and Steward, 1998b; Steward and Conway, 1998; Jones and Beckinsale, 1999). In support of this approach, Steward *et al.* (1994) argue that visual processing capabilities means that a graphic-based system is an effective way for humans to understand and manipulate relational data. While Conway and Overton (1994) emphasise the value of the network graphic, over textual and matrix displays, for its ability to encode a variety of quantitative and qualitative data in an efficient and effective manner.

The origin of the graphic representation of social networks may be traced back over 60 years to the 'sociogram' of Moreno (1934). Moreno devised the sociogram as a way of mapping social configurations to aid the researcher in visualising the channels through which one individual might be able to influence another, for example. The classic Hawthorne studies employed sociograms as a means to reveal the patterns of informal organisational relationships of managers and workers with regard to various aspects of group behaviour, such as

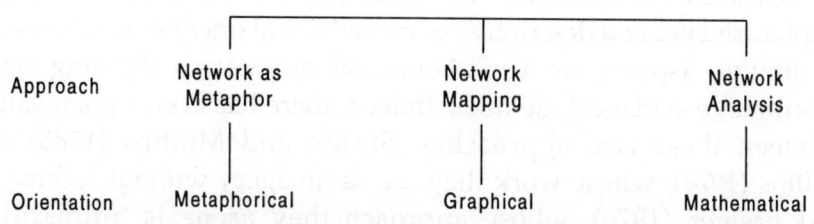

Fig. 1. Alternative approaches and orientations in studying networks (Conway, 1997b; Jones *et al.*, 1998).

involvement in job trading, helping, friendships and antagonisms (Roethlisberger and Dickson, 1939). The graphical mapping of informal networks was also a key aspect of a number of studies concerning the communication patterns of engineers during the early 1970s (Allen, 1970; 1977; Frost and Whitley, 1971). However, despite the power of the graphic, the dominant forms of network representation since this time have been textual narrative or mathematical display. Klovdahl (1985:313) has attributed this under-utilisation of the network graphic to "the time and tedium involved in producing hand-drawn diagrams" and "the impossibility of manipulating these [diagrams] once they are drawn". Yet recent advances in computer hardware, particularly in relation to the personal computer, and the availability of network graphics software now provide the network analyst with the tools to construct, manipulate, analyse and interpret network graphics with ease and speed. This paper places the network graphic 'centre stage', due to its importance in reducing, analysing, and displaying relational data. Furthermore, through the use of a set of snap-shots of the social interactions and linkages between individuals, the network graphic can be shown to be a powerful medium for illustrating the dynamic of social configurations.

While network maps aid the researcher in navigating the complex interactions and coalitions within organisations, it is worth reiterating the caution raised by Jones and Beckinsale (in their chapter), that the "map is not the territory".

Informal Organisation

Distinguishing Informal and Formal Organisation

Organisational charts and job descriptions generally reflect the formal structure or 'prescribed' network in a given organisation. Such prescriptions are often guided by the missions and strategies of the organisation (Chandler, 1962), even though their explicitness may vary greatly between one organisation and another. In contrast, informal or 'emergent' networks refer to the often covert and unsanctioned informal relations that emerge over and above such prescribed patterns of interaction (Roethlisberger and Dickson, 1939; Jacobson and Seashore, 1951; Blau, 1955; Mintzberg, 1973; Tichy, 1981; Monge and Eisenberg, 1987). The formal organisational structure is often referred to as the 'formal organisation' while the informal structure is frequently termed the 'informal organisation' or the 'social organisation' (Roethlisberger and Dickson, 1939; Allen, 1977; Stacey, 1996). Freeman (1991:503) argues that:

> "Behind every formal network, giving it the breath of life, are usually various informal networks.... Personal relationships of trust and confidence (and sometimes fear and obligation)....
> For this reason cultural factors such as language, educational background, regional loyalties, shared ideologies and experiences, and even common leisure interests, continue to play an important part in networking. An appreciation of these sociological factor...is a necessary complement to narrower *economic* explanations."

Blau and Scott (1962:6) posit that "informal organizations develop in response to the opportunities created and problems posed by their environment, and the formal organization constitutes the immediate environment of the groups within it". This view is supported by Tichy (1981:227) who argues that "metaphorically, a prescribed organizational network provides pegs from which emergent networks hang" and that "variations in a prescribed network therefore alter the emergent networks". However, research has indicated that other factors, such as proximity or expressive and affective needs (Moreno, 1934; Maisonneuve, 1952), also affect emergent networks. Indeed, Schwartz and Jacobson (1977:159) argue that while "to some extent these [informal] links...reflect the formal structure...the formal structure does not define the sociometric [informal] structure".

Table 1. A comparison of the characteristics of informal and formal organisation (adapted from Gray and Starke, 1984).

	Formal Organisation	Informal and Social Organisation
Structure a) origin b) rationale c) stability	Prescribed Rational Stable	Emergent Emotional Dynamic
Influence a) Base b) Type c) Flow	Position Authority Top-down	Personality Power Bottom-up
Communication a) Channels b) Networks	Formal channels Well-defined and follow formal channels.	'Grapevine' Poorly defined and cut across formal channels.
Individuals included	Those indicated by formal position and role.	Only those deemed 'acceptable'.
Basis for interaction	Prescribed by functional duties and position.	Spontaneous and personal characteristics.

Formal and informal organisation may be distinguished along a series of dimensions, including the origin and stability of the structure, the nature and flow of influence, and the basis for interaction of individuals in the network; these are summarised in Table 1 (adapted from Gray and Starke, 1984:412).

It has been argued that the degree to which the formal prescribed and the informal or social emergent networks overlap, is dependent upon whether the organisation is 'mechanistic' or 'organic' (Tichy *et al.*, 1979). The mechanistic organisation uses bureaucratic principles to design, plan and prescribe roles. Thus, the influence and information channels are likely to be highly prescribed. In addition, given high task certainty, individuals will have little discretion in their choice of work-related contacts. In contrast, organic organisations are most effective when there is high task uncertainty. Since complex or highly variable tasks cannot be pre-programmed, it is likely that informal relations arise to accomplish them (Burns and Stalker, 1961). Thus, Tichy *et al.* (1979:514) contend that:

> "This implies that the emergent network of interaction will closely follow the prescribed network in a mechanistic setting...[while] in organic organisations, emergent networks may differ considerably from prescribed networks."

Despite differences between the formal and informal organisation, Thomas (1994:17) notes: "...it is essential to remember that they are social constructs — and as such, their perpetuation is contingent on their repetition. In other words, social systems...are purposive entities that can be amended, dissolved, and reconstructed through the action of their participants." Furthermore, Whitley (1977:55) argues that "What an organisation 'is'...depends on the purpose of one's examination and an understanding of the numerous social institutions which, in varying ways and to varying extents, structure the reproduction of it as a social 'fact'."

Dis-Entangling Varying Notions of Informal and Social Organisation

It is apparent from a review of the organisational studies literature that variations exist in the meaning attached to the terms 'informal' or 'social' organisation; Mouzelis (1967) distinguishes between four categories of meaning: (1) informal as deviation from the formal; (2) informal as irrelevant to organizational goals; (3) informal as unanticipated; and (4) informal as "what really goes on in organisations". However, Burns and Flam (1987:232) argue that "organizational theorists have in large part failed to make systematic distinctions between different types and origins of informal rules...".

From the range of descriptions and definitions of 'informal organisation' that exist in the literature, it is possible to envisage a spectrum from the formally prescribed organisation chart, through to the 'purely' social groupings within and transcending the organisation. However, this spectrum may be more usefully presented as a simplified typology, as shown in Table 2. In developing a more fine-grained notion of the non-formal organisation, a distinction is made here between the informal organisation or network, in which task related interactions occur, and the social organisation or network, within which non-task related activities are undertaken. An intermediate level between the formal and informal structure is also sometimes recognised; this is termed the 'quasi-formal' structure, and refers to those relations that are built through the extensive use of flexible and transient teams and task-forces that are formally sanctioned by the organisation but are not reflected in the formal organisation chart (Schoonhoven and Jelinek, 1990).

Table 2. Distinguishing between formal and informal organisation structures.

Formal Organisation	Quasi-Formal Organisation	Informal Organisation	Social Organisation
Prescribed (Organisation Chart)	Emergent	Emergent	Emergent
Explicitly laid out and sanctioned by the Formal Organisation	Explicitly sanctioned by Formal Organisation.	Often explicitly encouraged through 'open' corporate cultures e.g., HPWay.	Often unrecognised, though sometimes encouraged.
Task Related	Task Related	Task Related	Non-Task Related

Accounting for the Emergence of Informal Organisation

Stacey (1996) argues that the need for informal organisation arises from two major failings of bureaucratic control: firstly, the subordination of individuality, and the alienating and de-motivating nature of bureaucracy; and secondly, the inability of the bureaucratic structure to handle environmental ambiguity and uncertainty. Stacey (1996:340) contends that:

> "People deal with the shortcoming of the bureaucratic system by colluding to operate a 'mock' bureaucracy (Gouldner, 1964) and acting instead within an informal organisation that they set up themselves (Blauner, 1964). The 'mock' bureaucracy is one in which all pay lip service to the rules but tacitly agree not to enforce them. The appearance of rationality and order is thus maintained, and any conflict that might have been generated by the applications of inappropriate rules is avoided."

Indeed, Tichy (1981:225) notes that "social networks play important roles in business organizations", since "unplanned structures... emerge because organizations are so complex that plans can never anticipate all contingencies"; Burns and Flam (1987) term this the "principle of bounded rationality". The importance of social or informal networks to innovative organisations is highlighted by Kreiner and Schultz (1993:189) who argue that:

> "In recent years, the traditional boundary activities bridging the company to its environment has been supplemented with a host of collaborative ventures.... Most noticeable, probably, is...when it takes the form of strategic alliances we are becoming aware that much of such collaboration is also pursued along unpaved paths in the undergrowth of less formalised, personalised networks."

Stacey (1996:341) sees the informal organisation as the mechanism that people employ to "deal with the highly complex, the ambiguous, the unpredictable, the inconsistent, the conflicting, the frustrating, and the alienating. They use it to satisfy social and motivational need...and as the tool to promote innovation and change". While Burns and Flam (1987) argue that organisations are "embedded in larger social structures and are partially open to the introduction of unofficial or informal rule systems"; this they term 'the principle of insufficient exclusivity'.

Organisations as Overlapping Sets of Networks and Rule Systems

A 'complex rule system' is one in which multiple (and often contradictory) rule systems are espoused by different groups. Burns and Flam (1987) term such social settings as 'polylithic'. Organisations have long been perceived in such a way; for example, Weber (1968:32)

> "The fact that, in the same social group, a plurality of con-
> tradictory systems of order may all be recognized as valid, is
> not a source of difficulty for the sociological approach. Indeed,
> it is even possible for the same individual to orient his action
> to *contradictory systems of order*. This can take place not only
> at different times, as is an everyday occurrence, but even in
> the case of the same concrete act."

These contradictory rule systems emerge, develop, and are mobilised through a plethora of overlapping social groupings, or networks. For instance, Burns and Flam (1987:233) argue that such social groupings in organisations may emerge around:

> "Class cleavages (owners/managers versus workers);
> cleavages between professional and occupational groups with
> differing functions, power and resource control as well as

social paradigms based on their training and their roles in the organisation."

In the social network literature, the degree to which individuals are linked by multiple role relations, such as friend, social club member and work colleague, is termed 'multiplexity'. It is contended that the greater the number of role relations (or strands) linking two actors, the stronger the linkage (Tichy *et al.*, 1979). In addition, Boissevain (1974:30) argues that:

> "There is a tendency for single-stranded relations to become many-stranded if they persist over time, and for many-stranded relations to be stronger than single-stranded ones, in the sense that one strand role reinforces another."

Laumann and Pappi (1976) argue that a social system may be differentiated in any number of ways, depending on the questions the researcher is interested in answering, but that certain bases of social differentiation are likely, in any empirical case, to be of special significance. Laumann and Pappi (1976:6) also contend: "... that there exists a multiplicity of social structures in any complex social system that arises out of the many possible types of social relationships linking positions [actors] to one another".

Earlier a distinction was made between formal, quasi-formal, informal and social organisation, based on the degree to which a 'social structure' is sanctioned and prescribed by management, and whether or not its activities are task related. Alternatively, such structures may be distinguished by the nature of what flows through particular linkages: this is termed 'interactional content' (Mitchell, 1969) or 'transaction content' (Fombrun, 1982) in the social network literature. Tichy *et al.* (1979) distinguish between affect, power, information and goods; networks constructed using this distinction are generally expressed in terms of the category of what is exchanged, hence friendship networks, influence networks, communication networks and economic networks, respectively. Tichy (1981:227) also argues that organisations are made up of "a multiplicity of networks arising out of many types of relationships, and each of these networks has its own structural and functional logic".

Krackhardt and Hanson (1993:111) identify three types of network: (1) the advice network: to reveal the prominent players in an organisation on whom others depend to solve problems and provide technical information; (2) the trust network: to reveal the pattern of sharing with regard to delicate political information, and support in a crisis; and (3) the communication network: to reveal the employees who talk to each other on a regular basis. In mapping out the various types of networks in the design group of a California-based computer company, Krackhardt and Hanson (1993) found that major differences were revealed in the key individuals and structures highlighted, between the formal, advice, communication and trust networks. They argue that an analysis of the trust and advice networks, in particular, provided a clearer picture of the dynamics at work in the design group they investigated. Stephenson and Krebs (1993) employed a similar typology in their study of employee diversity in a large US organisation. By mapping the prescribed, work, information, and support (mentoring) networks among a groups of employees and by distinguishing between gender and race, Stephenson and Krebs (1993) were able to highlight the degree of isolation/integration of female and minority employees.

Boundary-Spanning: A Key Characteristic of Informal and Social Organisation

A key characteristic of informal and social organisation is their tendency to span organisational boundaries: team boundaries, functional boundaries, and even the organisational boundary itself. Such boundary spanning interaction is the essence of the 'interactive model' of innovation (Rothwell and Zegveld, 1985). The interactive model places great emphasis on the ability of innovative organisations to manage relationships across interfaces, both within the firm (between project groups, functional departments, and divisions), and externally (within and across industrial sectors, geographical regions, and nations). In particular, studies of successful technological innovation have highlighted the importance of the internal marketing and R&D interface (Rothwell *et al.*, 1974, Calantone and Cooper, 1981; Bonnett, 1986) and of interacting with external organisations such as customers and suppliers (for example, Langrish *et al.*, 1972; Hippel, 1988; Conway, 1994).

Many studies have indicated the importance of informal or personal 'boundary-spanning' to the innovation process (Myers and Marquis, 1969; Utterback, 1971; Hippel, 1977; 1987; Allen *et al.*, 1983; Schrader, 1991; Kreiner and Schultz, 1991; Shaw, 1993; Conway, 1995; Steward and Conway, 1996) and particularly in relation to the transfer of 'tacit' knowledge (Senker, 1992; Senker and Faulkner, 1993). The importance of an appreciation of sociological factors in the understanding of personal boundary-spanning relationships is also recognised by others (for example, Lundvall, 1988; Bianchi and Bellini, 1991; Saxenian, 1991).

The Tensions Between Formal and Informal Organisation

Dissonance in Goals and Objectives

The informal organisation has been shown to be a valuable mechanism through which 'fresh' ideas and information filter into the innovation process, and as such, it represents an important 'intangible' organisational resource which is difficult for competitors to replicate. However, although there is much evidence to highlight the efficacy of the informal organisation in supplementing and complementing the activities of the formal organisation, "emergent [informal] networks can be dysfunctional as well as functional" (Tichy, 1981:225). This is not surprising, since as Burns and Flam (1987:214) note, "participants in the organization bring into it external statuses, relationships, network and organizational ties, each with their own social rule system, which may or may not contradict the formal system". Furthermore, while management may set the legal para-meters for informal exchange behaviour "what actually gets traded is determined by day-to-day interactions of engineers, marketers, and product developers" (Hamel *et al.*, 1989:136). Wolek and Griffith (1974:411) see this reliance on the informal organisation as "somewhat troublesome, for it is the *formal* channels which seem to be much more amenable to control and institutional support". Wolek and Griffith (1974:411) also note that informal networks are "sometimes interpreted as a sign of both weakness and need for better formal systems"; certainly informal networks present a number of stresses in attempting to steer the organisation towards formally defined goals and objectives.

A key managerial concern involves the flow of information across the organisational boundary. Informal boundary-spanning activity not only provides for the sourcing and acquisition of information and know-how, but can also result in information 'leakage'. Mansfield (1985) postulates that the rapid diffusion of technology via informal channels is one reason why many firms have difficulty in appropria-ting benefits from their innovations. This view is supported by Carter (1989:158), who argues that "exchangers of information do incur costs. The cost to the trader...is not the loss of the information itself, but rather the *competitive back-lash*". Thus, the information transfer behaviour of an employee cannot necessarily be assumed to be in accordance with the economic interests of his or her employer. This dissonance may arise where the 'trading' or 'sharing' of information by employees is guided by personal objectives, or even misguided, due to the insufficient availability of managerial information to enable well-informed decisions to be made. While Hamel *et al.* (1989) suggest measures to restrain informal boundary-spanning activity, Schrader argues that an organisation should employ mechanisms to induce desirable information transfer behaviour. This may include incentive schemes to motivate employees to act in the interests of the organisation and mechanisms to diffuse information internally.

The unpredictable nature of the linkages and interaction patterns within informal networks provides further consternation for managers in organisations. Carter (1989:155) argues that "because knowhow trading is informal and *off-the-books* such trading is difficult for the firm to evaluate and to manage". In addition, Kreiner and Schultz (1993), in their study of the Danish bio-technology sector, found that "the norms governing the interaction seem to reside in the network itself rather than in any of the participating organisations". The transient and intangible nature of informal organisation is highlighted by Mueller (1986:155), who sees informal networks as "short-lived, self-camouflaging and adisciplinary. They are invisible, uncountable, unpollable, and may be active or inactive".

Given the importance of informal boundary-spanning activity to the innovation process and the reliance on a relatively small number of specific individuals acting as 'boundary-spanners', the organisation is to some extent vulnerable. Indeed, Lawton-Smith *et al.* (1991:468) argue that "the downside of the key role which personal relationships play in collaborative ventures is over-dependence on certain individuals". However, in his study of the communication patterns of engineers, Allen (1977) found that gatekeepers were easily recognised by the organisation, with an overlap between guesses of the management and the study data of around 90%. In addressing this concern, Allen (1977) argues against formalising the role of boundary-spanners, which he believes "seems unnecessary and could even prove undesirable", favouring "recognition be afforded on a private, informal basis".

Competing Power Bases

Much of the innovation studies literature referred to in the previous section tends to see the informal organisation as misdirected but essentially benign, and thus in need of management and control.

This is perhaps not surprising, since as Knights and Murray (1994:31) note:

> "The process of OP [organisational politics], and the vast array of individual and collective strategies it involves, is necessarily riven through with a central paradox. For a great deal of managerial practice constructs a reality of its own activity that denies the political quality of that practice."

The same can be said for the social quality of organisational life.

The organisational studies literature stresses this more critical perspective; it locates 'politics' at the heart of informal organisation. Burns and Stalker (1994:188) argue that "no concern, it is safe to say, is without political or social conflict which generate, or contribute to, manifest inefficiencies of communication within the working organisation". For Stacey (1996:340), informal networks:

> "Are essentially political in nature; that is people handle conflicting interests through: persuasion and negotiation; implicit bargaining...and application of power that takes the form of influence rather than authority; that influence is derived from personal capability and the breadth of other network contacts...the informal group can often exert sanctions — the fear of rejection — that are stronger than those of the formal organisation."

Just as the informal organisation is in constant flux, so too is the nature and locus of conflict and 'challenge' within the organisation, both in relation to the formal organisation, and to other cliques and coalitions within the informal organisation. Indeed, for Burns and Flam (1987:214) "...power relations among the actors or groups advocating different [rule] systems become critical, since these will in part decide which of several competing or contradictory social rule systems will prevail." Burns and Flam (1987:233) also refer to the emergence of new power bases:

> "Informal status hierarchies and leadership patterns develop, countervailing and providing a basis of challenge to those formally designed and supervised by management."

Pettigrew (1973) argues that key to an understanding of the 'political landscape' of an organisation is an appreciation of access to, and control of, information, since these are essential sources of power. The informal organisation constitutes an important element of this landscape, since as Pfeffer (1981:130) argues: "Clearly, the power that comes from information control...derives largely from one's position in both the formal and informal communication networks". Research has highlighted a robust link between network centrality and the power accrued by individuals in organisations (Laumann and Pappi, 1976; Brass, 1984). Krackhardt (1990:343) argues that an important adjunct to network centrality is the 'accuracy' of an actor's perception of the informal organisation, or what Freeman *et al.* (1988) term 'social intelligence': "Power accrues not only to those who occupy central network positions in organizations but also to those who have an accurate perception of the network in which they are embedded". In fact, as Freeman *et al.* (1988) demonstrate, network centrality and social intelligence are closely inter-connected; they found that an actor's ability to accurately recall social structure was a function of whether they were a member of the core group or, a peripheral or transitory member.

Although frequently ignored, a number of recent studies have highlighted the central role of politics in the innovation process (Frost and Egri, 1990; Jones and Stevens, 1999; Jones and Beckinsale, 1999). Such conflict can be seen to originate in the competition for scarce resources, but also in the preservation or promotion of group or individual interests.

Methodology

Parsons (1951) argues that a social system may be differentiated in any number of ways depending on the questions that the analyst is interested in answering. However, it is also true that certain bases of social differentiation are likely to be of special salience and significance in a given empirical case (Blau, 1975). Furthermore, Mitchell (1969) stresses that the network researcher must always select particular elements of the 'total network' for attention; what he terms 'partial networks'. He conceptualised the total network of a society as "the general ever-ramifying, ever-reticulating set of linkages that stretches within and beyond the confines of any community or organisation" (Mitchell, 1969:12). This process of selecting partial networks for further study is commonly termed abstraction (Scott, 1991). There are two key decisions to be taken in abstraction: the first focuses on the rules of inclusion based on the attributes of the actors or links, and/or participation of the actors themselves in some activity or exchange, termed the 'definitional focus' (Laumann *et al.*, 1983); the second relates to the manner in which the abstraction is 'anchored' or 'centred' which may be around a particular actor or group of actors. This may be termed 'nodal-anchoring' (Conway, 1994).

The following short case-study adopts a 'socio-centred action-set' approach. This is useful for two reasons: firstly it removes the need to decide on some arbitrary boundary to be drawn around those to be included (a particular problem with 'ego-centred' networks), and secondly, as Aldrich and Whetten (1981:386) note, the "focusing on resource flows or boundary-role interaction avoids the problem of mixing transitory or ephemeral relations with enduring or consequential ones".

The case-study presented in this chapter is based upon material collected during an MBA dissertation supervised by the author (Dass, 1997). This information has since been supplemented through a further interview and use of documentary evidence. The MBA student had worked for the organisation under investigation for a number of

Table 3. Summary of alternative options for the abstraction of partial networks (Conway, 1994; Conway and Steward, 1998).

Definitional Focus		Nodal-Anchoring	
Network Type	Description	Network Type	Description
Attribute Network	Inclusion rule based on specified attributes of actors or links.	Socio-Centred Network	Focus of analysis centred on group of actors within network.
Transaction Network	Inclusion rule based on actor participation in specified exchange.	Ego-Centred or Focal Network	Focus of analysis centred around single actor within network.
Action-Set	Inclusion rule based on actor participation in specified event.		

years, and this allowed him to recall relevant case-studies illustrating the value of the informal organisation, and to gain access to the individuals involved. For each of the case-studies, data was collected through open interviews with the relevant individuals. This data included information regarding the links and interactions each individual had with others in the organisation with regard to the specified case-study.

Introduction to the Case-Study

The Organisation

The organisation is a Housing Trust. It has a hierarchical organisational structure, which reflects the geographical spread of its activities.

The Head Office incorporates a number of centralised functional departments, such as Finance and Personnel. At the time of this case study there were five Regional Offices, each had a Regional Manager who was responsible for overseeing the various Housing Estates Offices located within their region. Each Region was further divided into Areas, each incorporating four or five Estate Offices; the day-to-day activities of these were overseen by an Area Manager. The appointment of a new Chief Executive Officer in 1992 marked the starting point for the remoulding of the Trust. As part of this process, the Trustees initiated a review of the Trust's IT requirements in August 1993, aimed at developing an IT strategy that would take the organisation into the next decade.

The following case-study reaffirms the importance of informal networks, in this instance in overcoming technical problems faced by the organisation in the implementation of an IT system; it also highlights the tension between the formal and informal organisation. Although the formal organisation stood to benefit greatly from the mobilisation of the informal network, these interactions were perceived as undermining the legitimate power and personal credibility of key actors in the formal organisation. As a consequence, informal activity was marginalised, and to a degree 'criminalised', as the formal organisation sought to reassert the status of the formal roles, interactions and activities.

Background: Formulating an IT Strategy and Selecting the Pilot Housing Estate

From the mid-1980s, the Trust had been using a central computer located within the Finance Department at Head Office. This centralised computer system already dealt with the processing of rents and accounts, but had limited functionality and was reaching the end of its useful life. Furthermore, up until the time of the IT review, personal computers had been restricted to Head Office and the five Regional Offices. It was increasingly felt that the installation of computers within the individual Estate Offices would provide greater autonomy to Housing Estate Managers and allow them to make more informed decisions.

In August 1993, the Deputy Accountant at the Head Office was charged with setting up a Computer Working Party in order to direct and oversee the review of existing computer facilities and future computer needs. In fact, one of the main issues that arose from the review was the lack of representation of Estate Office based staff in the Working Party. To address this, it was decided to include one Estate Manager, and that the person chosen would be selected from the Region that would act as the pilot for the installation of the new IT system. The Midlands Region was chosen for the pilot; this had been expected since the region had always been the testing ground for the implementation of new policies and procedures. However, the choice of Estate Manager was a surprise. It was commonly agreed among the Estate Managers in the Midland Region that Barry from the Bancroft Estate Office was the ideal candidate, he was, after all, a 'wizard with computers'. In the event, Colin was selected by Head Office to act as the the representative on the Computer Working Party. Colin had served in the Armed Forces and had a reputation as a strict and officious man who never questioned his superiors. These qualities made it difficult for staff to talk to Colin on an informal basis. Furthermore, his knowledge of computers was very limited. Barry felt let down.

The Case-Study: Pilot Implementation of an IT System

Initiating the Pilot Implementation

The following month an 'invitation to tender' was announced and a short-list of computer companies was whittled down to one. The winner of the contract was charged with supplying and installing the full system, incorporating both computer hardware and software.

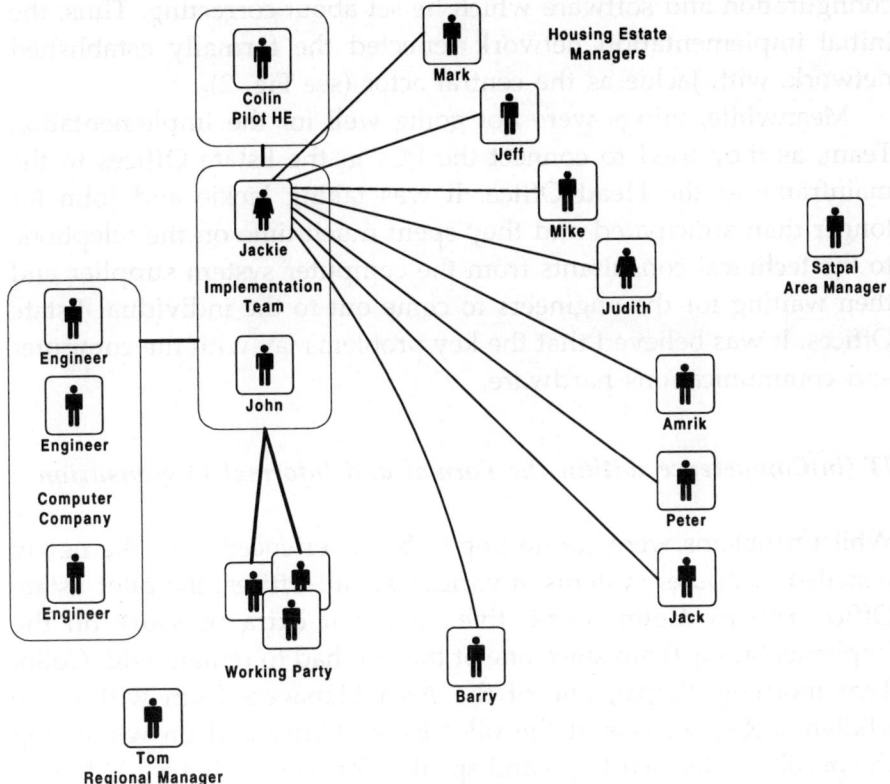

Fig. 2. The early formal network established by the implementation team.

Towards the latter part of 1994, the Trust saw the first of the computers arriving at the Estate Offices in the Midlands Region. However, the Estate Managers were immediately sent a memo instructing them not to open the boxes and attempt to set up the computers; instead the Implementation Team would travel to each Estate Office and unpack, set-up and check the computers; this team included Jackie (the main point of contact for the pilot) and John (the more senior of the two). Barry ignored the memo, set-up his own computer and immediately identified a number of faults with the computer configuration and software which he set about correcting. Thus, the initial implementation network reflected the formally established network, with Jackie as the central actor (see Fig. 2).

Meanwhile, things were not going well for the Implementation Team, as they tried to connect the PCs in the Estate Offices to the mainframe at the Head Office. It was taking Jackie and John far longer than anticipated and they spent much time on the telephone to the technical consultants from the computer system supplier and then waiting for the engineers to come out to the individual Estate Offices. It was believed that the key problems lay with the computer and communications hardware.

IT (in)Competence within the Formal and Informal Organisation

Whilst problems were continuing to be experienced with the newly installed computer systems at various Estate Offices, the pilot Estate Office, run by Colin, went 'live'; this put extra pressure on the Implementation Team since one of the two had to remain with Colin. That morning, Satpal, one of the Area Managers from within the Midlands Region, was at the pilot Estate Office and on witnessing the problems decided to go and speak informally to Barry. Although Barry's Estate Office was not 'live', he had managed to get his computer system fully operational. Barry explained to Satpal that the problems lay not with the hardware itself but with the software and the way the hardware had been configured; these problems were decreasing memory size and causing the computers to crash. The situation was compounded further by the Implementation Team who were 'bodging' rather than solving the problems. A few days passed and Satpal learnt that the date for the whole of the Midlands Region to go live had been postponed indefinitely, until the computer problems had been resolved at the various Estate Offices. Satpal approached Tom, the Midlands Regional Manager, and explained that Barry had solved the technical problems. He was impressed and went to speak to Jackie to pass on the information.

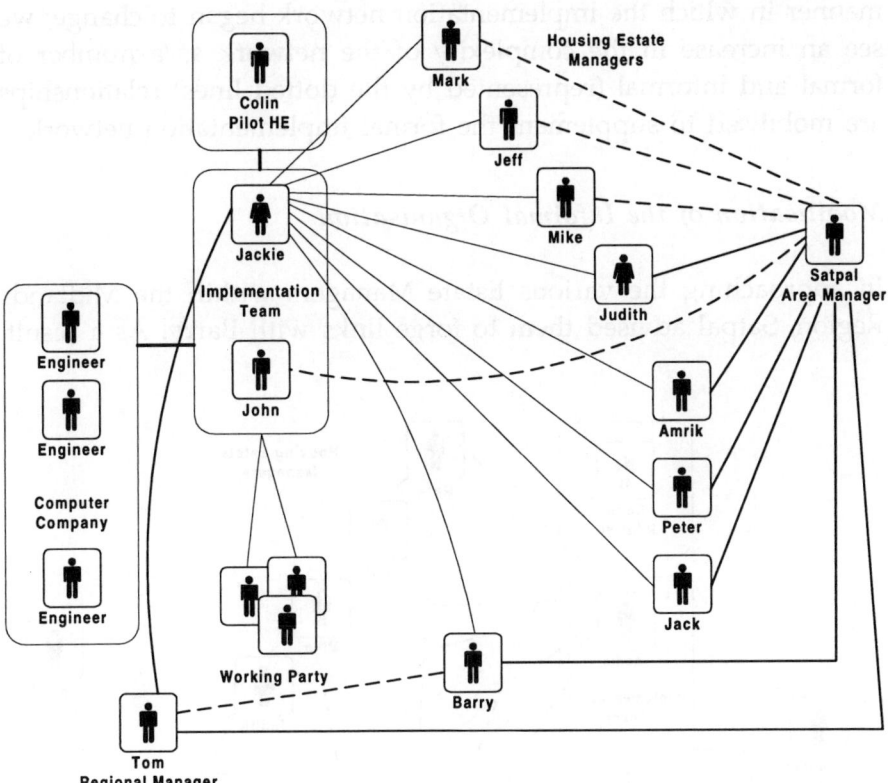

Fig. 3. The emergence of Satpal as a boundary-spanner.

Jackie was extremely embarrassed and felt her authority had been undermined.

Following his meeting with Barry, Satpal went to speak formally with those Estate Managers within his area of authority, and informally with those outside of his domain, about the problems they were experiencing and the potential solutions. As a result, a new informal 'hub' began to emerge in the network, as Satpal acted as informal facilitator and intermediary between Barry and the other Estate Managers in the Midlands Region. Figure 3 illustrates the manner in which the implementation network began to change; we see an increase in the complexity of the network as a number of formal and informal (represented by the dotted lines) relationships are mobilised to supplement the formal implementation network.

Mobilisation of the Informal Organisation

In approaching the various Estate Managers within the Midlands Region, Satpal advised them to forge links with Barry. As a result,

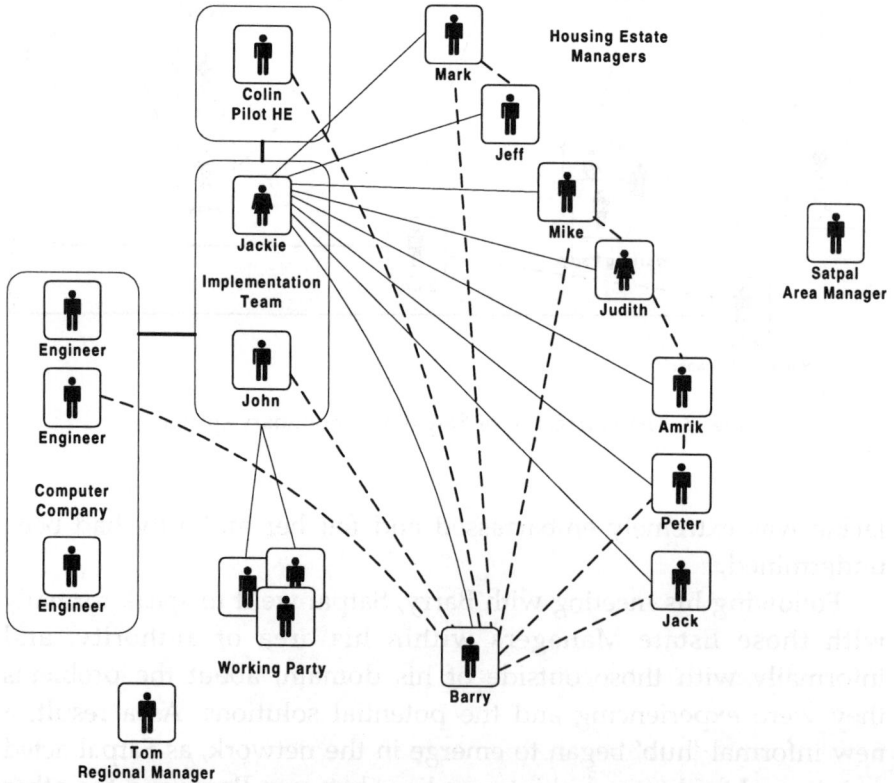

Fig. 4. The emergence of the informal network as an alternative to the formal network.

Jackie and John began to receive far fewer telephone enquires, while Barry started to speak informally to a number of the Estate Managers to advise them how to get their computer systems up and running correctly. Once again a new informal 'hub' emerged in the network, as Satpal moved to the 'sidelines' and Barry began to interact with the various Estate Managers in the pilot implementation. The impact of these interactions on the morphology of the network is illustrated in Fig. 4; here we see an informal network emerging in direct 'competition' with the formal network, as the source of technical expertise and support.

Reassertion of the Formal Organisation

Meanwhile schemes were being hatched. To hide the embarrassment that had arisen from Barry's informal problem-solving, Jackie spoke to John, emphasising not so much that Barry had solved the technical problems, but to highlight the fact that he had breached the strict instructions sent by Head Office "that Estate Managers must not tamper with the computer equipment". She further pointed out that not only had Barry set-up his own computer, but he had also modified the software, stressing that this ought to be a disciplinary offence. John visited Barry and spent the day learning the nature of the problems with the computer system and the solutions Barry had employed to rectify them. A week later, Barry received a letter from Head Office stating that any further tampering with the computer system by Estate Office staff would result in disciplinary action. This in effect reasserted the position and influence of the formal network and Jackie once again became the hub. Meanwhile, the informal interactions dwindled and Barry became marginalised. The impact of the reassertion of the formal implementation network, and of Jackie as the 'hub', can be seen in Fig. 5.

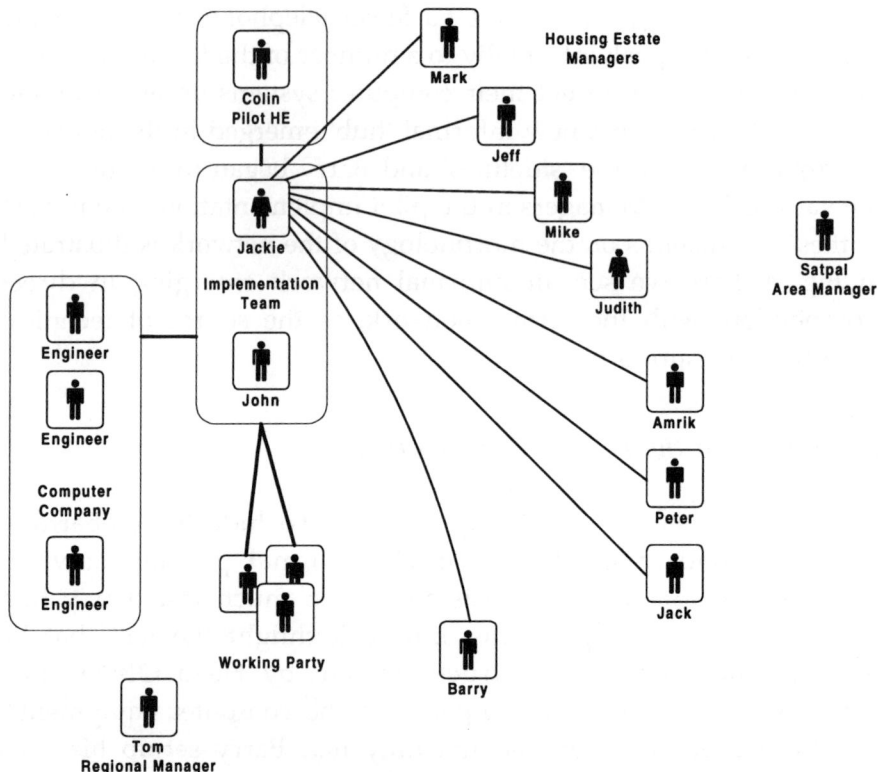

Fig. 5. The reassertion of formal network and the suppression of the informal network.

Within a short period of time, the computer problems at the various Estate Office had been rectified and the Midlands Region successfully went 'live'. Both Jackie and John were subsequently promoted. The Implementation Team were openly congratulated and thanked for their efforts by Head Office. However, Barry did not receive any recognition or thanks for his efforts and innovativeness in solving the technical problems that threatened to throw the whole pilot IT implementation into disarray.

What Does the Case-Study Help to Illustrate ?

The case-study serves to reinforce a number of the points raised through the literature review in earlier sections of this chapter. Firstly, the case highlights the potential for tension between the formal and informal organisation; here the emergent informal network is viewed as a competing power-base by key actors within the formal implementation network; the two networks have alternative power sources: the formal — legitimate power; the informal — power derived through knowledge. In attempting to reassert the formal organisation, the legitimate power of the formal implementation network is mobilised; this effectively dismantles the informal element of the action-set. Secondly, the network maps, representing snap-shots of the action-set at different time-frames, illustrate the dynamic nature of the morphology of the action-set, and in particular the informal network. Furthermore, the 'ebb and flow' of informal interactions appear to occur within the constraints set by the formal organisation. Thirdly, the network maps also help to identify key actors or 'hubs' within the network. Viewing these network maps over time helps to highlight the shifts in power and influence represented within the network as shifts in the 'hub' position from one actor, or group of actors, to another. Finally, the case is intended to demonstrate the utility of the network mapping approach for focusing attention on the linkages and interactions between actors, and in the power of the graphic in depicting relational data.

Concluding Comments

This chapter has set out to explore the nature of informal organisation and its relationship with the formal organisation. It outlines an approach to researching informal organisation, termed 'network mapping'; this may be positioned between the extremes of the metaphorical and mathematical network orientations that have dominated the network literature. The 'network mapping' approach is employed in a short case-study of the implementation of an IT system in a Housing Trust, to highlight both the dynamism and tension between the formal and informal network, as well as the utility of the network graphic.

By and large, the innovation literature espouses the virtues of informal networking in the innovation and knowledge building process. Informal networks are viewed as an important structure for promoting communication, integration, flexlibility, and novelty, within and between organisations. They are seen as monolithic structures that supplement, complement and add value to the formal organi-sation. On the whole, the innovation literature ignores the dysfunctional potential of the informal or social organisation. The organisation studies literature is more critical, viewing informal or social networks as a key element of the political landscape of organisations; it views the informal or social organisation as polylithic, as a complex, dynamic set of overlapping cliques. These cliques have divergent goals and objectives, and represent competing power bases within the organisation. This later perspective of the role and nature of informal networks is crucial in developing an understanding of the importance of informal interactions and structures in the in-novation process, and in the workings of organisations in general. This supports the view of Krackhardt and Hanson (1993:111), who argue that:

"Experienced network managers who can use maps to identify, leverage, and revamp informal networks will become increasingly valuable as companies continue to flatten and rely on teams. As organisations abandon hierarchical struc-tures, managers will have to rely less on authority inherent in their title and more on their relationships with players in their informal networks. They will need to focus less on over-seeing employees 'below' them and more on managing people across functions and disciplines. Understanding relationships will be the key to managerial success."

References

Alba, R. (1982) "Taking stock of network analysis: A decade's results," in *Research in the Sociology of Organizations: A Research Annual*, Vol. 1, S. Bacharach (ed.), JAI Press, Connecticut, 39–74.

Allen, T. (1970) "Communication networks in R&D laboratories," *R&D Management*, **1**(1), 14–21.

Allen, T. (1977) *Managing the Flow of Technology: Technology Transfer and the Dissemination of Technological Information within the R&D Organization*. MIT Press, Cambridge, Massachusetts.

Allen, T., Hyman, D. and Pinckney, D. (1983) "Transferring technology to the small manufacturing firm: A study of technology transfer in three countries," *Research Policy*, **12**(2), 199–211.

Auster, E. (1990) "The interorganizational environment: Network theory, tools, and applications," in *Technology Transfer: A Communication Perspective*, F. Williams and D. Gibson (eds.), Sage Publications, 63–89.

Barnard, C. (1938) *The Functions of the Executive*. Harvard University Press, Cambridge, MA.

Bianchi, P. and Bellini, N. (1991) "Public policies for local networks of innovators," *Research Policy*, **20**(5), 487–497.

Blau, P. (1955) *The Dynamics of Bureaucracy*. University of Chicago Press, Chicago.

Blau, P. (1982) "Structural sociology and network analysis: An overview, in *Social Structure and Network Analysis*, P. Marsden and N. Lin (eds.), Sage, Beverly Hills, 273–297.

Blau, P. and Scott, W. (1962) *Formal Organizations: A Comparative Approach*. Chandler, San Francisco.

Blauner, R. (1964) *Alienation and Freedom: The Factory Worker and His Industry*. University of Chicago Press, Chicago.

Boissevain, J. (1974) *Friends of Friends: Networks, Manipulators and Coalitions*. Basil Blackwell, Oxford.

Bonnett, D. (1986) "Nature of the R&D/marketing co-operation in the design of technologically advanced new industrial products," *R&D Management*, 16(2), 117–126.

Boorman, Scott and White, Harrison (1976) "Social structure from multiple networks: II. Role structures," *American Journal of Sociology*, 81(6), 1384–1446.

Boulding, K. (1985) *The World as a Total System*. Sage, Beverly Hills, California.

Brass, D. (1984) "Being in the right place: A structural analysis of individual influence in an organization," *Administrative Science Quarterly*, 29, 518–539.

Breiger, R. (1976) "Career attributes and network structure: A blockmodel study of a biomedical research network," *American Sociological Review*, 47, 117–135.

Burns and Flam (1987) *The Shaping of Social Organization*. Sage, London.

Burns, T. and Stalker, G. (1961) *The Management of Innovation*. Tavistock Publications, London.

Burns, T. and Stalker, G. (1994) *The Management of Innovation*. Oxford University Press, Oxford.

Burt, R. (1980) "Models of network structure," in *Annual Review of Sociology*, A. Inkeles (ed.), Vol. 6, 79–141.

Calantone, R. and Cooper, R. (1981) "New product scenarios: Prospects for success," *Journal of Marketing*, 45, 48–60.

Carroll, G. and Teo, A. (1996) "On the social networks of managers," *Academy of Management Journal*, 39(2), 421–440.

Carter, A. (1989) **"Knowhow trading as economic exchange,"** *Research Policy*, 181(2), 155–163.

Chandler, A. (1962) *Strategy and Structure*. MIT Press, Cambridge, Massachusetts.

Collins, H. (1974) "The TEA set: Tacit knowledge and scientific networks," *Science Studies*, 4, 165–185.

Conway, S. (1994) "Informal boundary-spanning links and networks in successful technological innovation," unpublished Ph.D. dissertation, Aston Business School, Birmingham.

Conway, S. (1995) "Informal boundary-spanning networks in successful technological innovation," *Technology Analysis & Strategic Management*, **7**(3), 327–342.

Conway, S. (1997a) "Strategic personal links in successful innovation: Link-pins, bridges, and liaisons," *Creativity and Innovation Management*, **6**(4), 226–233.

Conway, S. (1997b) "Focal innovation action-sets: A methodological approach for mapping innovation networks," Research Paper Series, No. RP9702, Aston Business School Research Institute, Birmingham.

Conway, S. and Overton, M. (1994) "Constructing the network graphic: A palette of options," Doctoral Working Paper Series, No. 15 (NS), Aston Business School, Birmingham.

Conway, S. and Steward, F. (1998a) "Networks and interfaces in environmental innovation: A comparative study in the UK and Germany," *Journal of High Technology Management Research*, **9**(2), 239–253.

Conway, S. and Steward, F. (1998b) "Mapping innovation networks," *International Journal of Innovation Management*, **2**(2), 165–196.

Dass, S. (1997) "How the informal network adds value to the formal structure of an organisation," unpublished MBA dissertation, Aston Business School, Birmingham.

Fombrun, C. (1982) "Strategies for network research in organisations," *Academy of Management Review*, **7**(2), 280–291.

Freeman, C. (1991) "Networks of innovators: A synthesis of research issues," *Research Policy*, **20**(5), 499–514.

Freeman, L., Freeman, S. and Michaelson, A. (1988) "On human social intelligence," *Journal of Social and Biological Structures*, **11**, 415–425.

Frost, P. and Egri, C. (1990) "Influence of political action on innovation: Part II," *Leadership and Organizational Development Journal*, **11**(2), 4–11.

Frost, P. and Whitley, R. (1971) "Communication patterns in a research laboratory," *R&D Management*, **1**(2), 71–79.

Galaskiewicz, J. (1979) *Exchange Networks and Community Politics*. Sage, Beverly Hills.

Gouldner, A. (1964) *Patterns of Industrial Bureaucracy*. The Free Press, New York.

Gray, J. and Starke, F. (1984) *Organizational Behaviour: Concepts and Applications*. 3rd ed., Charles Merrill, Columbus, Ohio.

Hamel, G., Doz, Y. and Prahalad, C. (1989) "Collaborate with your competitors—and win," *Harvard Business Review*, **67**(1), 133–139.

Hippel, E. von (1977) "Transferring process equipment innovations from user-innovators to equipment manufacturing firms," *R&D Management*, **8**(1), 13–22.

Hippel, E. von (1987) "Cooperation between rivals: Informal know-how trading," *Research Policy*, **6**(6), 291–302.

Hippel, E. von (1988) *The Sources of Innovation*. Oxford University Press, London.

Ibarra, H. (1993) "Network centrality, power and innovation involvement," *Academy of Management Journal*, **36**, 471–501.

Jacobson, E. and Seashore, S. (1951) "Communication practices in complex organizations," *Journal of Social Issues*, **7**, 28–40.

Jones, O., Conway, S. and Steward, F. (1998) "Introduction: Social interaction and innovation networks," *International Journal of Innovation Management*, **2**(2), 123–136.

Jones, O. and Beckinsale, M. (1999) "Analysing the innovation process: Networks, micropolitics and structural change," Research Paper Series, Aston Business School Research Institute.

Jones, O. and Stevens, G. (1999) "Evaluating failure in the innovation process: The micropolitics of new product development," *R&D Management*, **29**(2), 167–178.

Kanter, R. (1985) *The Change Masters*. Unwin, London.

Katz, D. and Kahn, R. (1966) *The Social Psychology of Organisations*. Wiley, New York.

Klovdahl, A. (1985) "Social networks and the spread of infectious diseases: The AIDS example," *Social Science of Medicine*, **12**, 1203–1216.

Knights, D. and Murray, F. (1994) *Managers Divided: Planning, Implementation and Control*, 8th Ed. (1st Ed. 1967), Wiley, Chichester.

Krackhardt, D. (1990) "Assessing the political landscape: Structure, cognition, and power in organizations," *Administrative Science Quartley*, **35**, 342–369.

Krackhardt, D. and Hanson, J. (1993) "Informal networks: The company behind the chart," *Harvard Business Review*, July/August, 104–111.

Kreiner, K. and Schultz, M. (1993) "Informal collaboration in R&D: The formation of networks across organizations," *Organization Studies*, **14**(2), 189–209.

Langrish, J., Gibbons, M., Evans, W. and Jevons, F. (1972) *Wealth From Knowledge: A Study of Innovation in Industry*. MacMillan, London.

Laumann, E. and Pappi, F. (1976) *Networks of Collective Action: A Perspective on Community Influence Systems*. Academic Press, New York.

Lawton-Smith, H., Dickson, K. and Smith, S. (1991) "There are two sides to every story: Innovation and collaboration within networks," *Research Policy*, **20**, 457–468.

Litterer, J. (1973) *The Analysis of Organisations*. 2nd ed.

Lundvall, B. (1988) "Innovation as an interactive process: From user-producer interaction to the national system of innovation," in *Technical Change and Economic Theory*, G. Dosi *et al.* (ed.), Pinter, London.

Lundvall, B. (1992) "Introduction," in *National Systems of Innovation: Towards a Theory of Innovations and Interactive Learning*, B. Lundvall (ed.), Pinter Publishers, London.

Maisonneuve, J. (1952) "Selective choices and propinquity," *Sociometry*, **15**, 123–134.

Mansfield, E. (1985) "How rapidly does new industrial technology leak out?" *Journal of Industrial Economics*, **34**(2), 217–223.

Marsden, P. and Lin, N. (1982) "Introduction," in *Social Structure and Network Analysis*, P. Marsden and N. Lin (eds.), Sage, Beverly Hills, California.

Mintzberg, H. (1973) *The Nature of Managerial Work*. Harper and Row, New York.

Mintzberg, H. (1979) *The Structuring of Organisations.* Prentice Hall, Englewood, New Jersey.

Mitchell, J. (1969) *Social Networks in Urban Situations.* Manchester University Press.

Monge, P. and Eisenberg, E. (1987) "Emergent communication networks," in *Handbook of Organisational Communication,* F. Jablin, L. Putman, K. Roberts and L. Porter (eds.), Sage Publications, London.

Moreno, J. (1934) "Who shall survive?: A new approach to the problem of human interactions," Nervous and Mental Disease Monograph Series, No. 58, Nervous and Mental Disease Publishing Company, Washington.

Moreno, J. (1953) *Who Shall Survive?: Foundations of Sociometry, Group Psychotheraphy and Sociodrama.* Beacon House, New York.

Morgan, G. (1986) *Images of Organization.* Sage, Beverly Hills, California.

Mouzelis, N. (1967) *Organization and Bureaucracy.* Aldine, Chicago.

Mueller, R. (1986) *Corporate Networking: Building Channels for Information and Influence.* The Free Press, New York.

Myers, S. and Marquis, D. (1969) *Successful Commercial Innovations.* National Science Foundation, Washington D.C.

Peppard, J. and Davis, P. (1996) "Using social network analysis to understand IT impacts in organisations," paper presented at the *BAM Conference,* Birmingham, September.

Pettigrew, A. (1973) *The Politics of Organizational Decision-Making.* Tavistock, London.

Pfeffer, J. (1981) *Power in Organizations.* Pitman, Boston.

Roethlisberger F. and Dickson, W. (1939) *Management and the Worker.* Harvard University Press, Cambridge, MA.

Rogers, E. (1982) "Information exchange and technological change," In *The Transfer and Utilisation of Technical Knowledge,* D. Sahal (ed.), Lexington Books, Lexington, Massachusetts, 105–123.

Rogers, E. (1987) "Progress, problems and prospects for network research: Investigating relationships in the age of electronic communication," paper presented at the *VII Sunbelt Social Networks Conference*, Florida, 12–15 February.

Rothwell, R. and Zegveld, W. (1985) *Reindustrialisation and Technology*. Longman, London.

Rothwell, R., Freeman, C., Horsley, A., Jervis, P., Robertson, A. and Townsend, J. (1974) "SAPPHO updated—project SAPPHO phase II," *Research Policy*, 3(3), 258–291.

Saxenian, A. (1991) "The origins and dynamics of production networks in the silicon valley," *Research Policy*, 20(5), 423–437.

Schoonhoven, C. and Jelinek, M. (1990) "Dynamic tensions in innovative firms: Managing rapid technological change through organisational structure," in *Managing Complexity in High Technology Organisations*, M. von Glinow and A. Mohrman (eds.), Oxford University Press, New York, 90–118.

Schrader, S. (1991) "Informal technology transfer between firms: Cooperation through information trading," *Research Policy*, 20(2), 153–170.

Schwartz, D. and Jacobson, E. (1977) "Organizational communication network analysis: The liaison communication role," *Organizational Behaviour and Human Performance*, 18, 158–174.

Scott, J. (1991) *Social Network Analysis: A Handbook*. Sage, London.

Senker, J. (1992) "The contribution of tacit knowledge to innovation," paper presented at the *Exploring Expertise Workshop*, University of Edinburgh, November.

Senker, J. and Faulkner, W. (1993) "Networks, tacit knowledge and innovation," paper presented at the *2nd ASEAT International Conference*, Manchester, 21–23 April.

Shaw, B. (1993) "Formal and informal networks in the UK medical equipment industry," *Technovation*, 13(6), 349–365.

Shrum, W. (1990) "Status incongruence among boundary-spanners: Structure, exchange, and conflict," *American Sociological Review*, 55(4), 496–511.

Shrum, W. and Mullins, N. (1988) "Network analysis in the study of science and technology," in *Handbook of Quantitative Studies of Science and Technology*, A. van Rann (ed.), North-Holland, Amsterdam.

Simmel, G. (1955) *Conflict and the Web of Group-Affiliations*. (Translated by K. Wolff and R. Bendix), Free Press, New York.

Stacey, R. (1996) *Strategic Management and Organisational Dynamics*. 2nd ed., Pitman Publishing, London.

Stephenson, K. and Krebs, V. (1993) "A more accurate way to measure diversity," *Personnel Journal*, 66–74.

Steward, F. and Conway, S. (1996) "Informal Networks in the origination of successful innovations," in *Technological Collaboration: The Dynamics of Cooperation in Industrial Innovation*, R. Coombs, P. Saviotti, A. Richards and V. Walsh (eds.), Edward Elgar, 201–221.

Steward, F. and Conway, S. (1998) "Situating discourse in environmental innovation networks," *Organization*, 5(4), 483–506.

Steward, F., Conway, S. and Overton, M. (1994) "Depicting innovation networks," in *Designs, Networks and Strategies*, Vol. 1, A. Francis, S. Horte and J. Pedersen (eds.), Commission of the European Communities, 185–201.

Thomas, R. (1994) *What Machines Can't Do: Politics and Technology in the Industrial Enterprise*. University of California Press, Berkeley.

Tichy, N. (1981) "Networks in organizations," in *Handbook of Organizational Design*, Vol. 2, P. Nystrom and W. Starbuck (eds.), Oxford University Press, New York, 225–247.

Tichy, N., Tushman, M. and Fombrun, C. (1979) "Social network analysis for organisations," *Academy of Management Review*, 4(4), 507-519.

Utterback, J. (1971) "The process of innovation: A study of the origination and development of ideas for scientific instruments," *IEEE Transactions on Engineering Management*, **EM-18**(4), 124–131.

Weber, M. (1968) *Economy and Society*. Bedminister Press, New York.

Weick, K. (1969) *The Social Psychology of Organising*. Addison-Wesley, London.

Wellman, B. (1983) "Network analysis: Some basic principles," in *Sociological Theory*, R. Collins (ed.), Jossey-Bass, San Francisco.

White, H., Boorman, S. and Breiger, R. (1976), "Social structure from multiple networks: I. Blockmodels of roles and positions," *American Journal of Sociology*, 81(4), 730–780.

Whitley, R. (1977) "Concepts of organization and power in the study of organizations," *Personnel Review*, 6(1), 54–59.

Wolek, F. and Griffith, B. (1974) "Policy and informal communication in applied science and technology," *Science Studies*, 4, 411–420.

Chapter 4

An Economic Perspective On Innovation Networks

David Parker[*] and Kirit Vaidya[†]

Introduction

Industrial associations, professional societies, joint ventures, strategic alliances, partnership sourcing and 'clustering' in industrial districts are all examples of networking among individuals and organisations. Sydow (1998:33) defines a network relationship 'as a long-term institutional arrangement amongst distinct but related organizations'. These arrangements are more loose than horizontal or vertical integration as a means of getting 'access to external resources, necessary in the pursuit of...opportunities' (Jarillo, 1988:39; for similar views, see Cook and Emerson, 1978; Swan and Newell, 1995:849; Conway and Steward, 1996).

According to Cravens, Piercy and Shipp (1998:204–206), networks have developed to provide flexibility in the face of major global change, to develop skills and resources, and to achieve operating economies. Inter-organisational networking has been interpreted as

[*]E-mail: d.parker1@aston.ac.uk
[†]E-mail: k.g.vaidya@aston.ac.uk

a key feature of the diffusion of new ideas across organisations (Kanter, 1990; Nohria and Eccles (eds.), 1992; Rogers, 1995; also see Cooke and Morgan, 1998:13; Porter, 1990 and Conway, 1995).[1] According to evidence examined by Hagedoorn (1993), the goals of most strategic alliance partners have been to gain access to new and complementary technologies, to speed up innovatory or learning processes and to upgrade particular activities such as research and development, marketing and distribution and manufacturing.

In the context of innovation, a network can be defined as the web of linkages between firms and other institutions to enable actors to benefit from information on opportunities and threats, acquisition of technical and market knowledge and to gain better access to inputs. Firms learn both from their own experience of design, development, production and marketing and from external sources which include their customers, their suppliers, their contractors (a particularly important part of Japanese firm behaviour), universities, government laboratories and agencies, consultants, licensors, licensees and competitors (through informal contacts and reverse engineering) (Freeman, 1998 and Nelson, 1993). The precise pattern of external and internal learning varies with the size of the firm, but many firms make use of external sources. For example, one study found that in Europe around 40% of companies performing some R&D had co-operated with partners external to the firm and such networking accounts for around 10% of total R&D budgets (Kleinknecht and Reyner, 1992; cited in de Laat, 1997:146). A number of studies cited by Freeman (1998) indicate that the frequency and intensity of the interaction with actual and potential users is a major determinant of innovative success.

In this chapter, the subject of networks and their role in innovation is appraised from an economic perspective.[2] The chapter begins with

---

[1]The same can apply to intra-organisational networking, in terms of diffusing new ideas within firms (Drucker, 1992; Hislop, *et al.*, 1997:432).

[2]The coverage is primarily based on recent developments in institutional and evolutionary economics. Space does not permit a detailed critique of the neoclassical approach and the Austrian School, though brief reference is made to them in Sec. 4.

some initial reflections on features of knowledge and innovation, which suggest an important role for external relationships in innovation within firms. The discussion then turns to the transaction cost literature which provides a useful departure point for the study of market and non-market arrangements for transacting (including networks). The evolutionary and resource-based theories of the firm, which are arguably more appropriate for the study of innovation, are then introduced to provide a more dynamic model of the firm and its external linkages. Discussion then turns to the role of trust followed by implications of spillovers of innovation in network relationships. A simple game theory model is used to analyse the rationale for network relationships.

The economic perspective seeks explanations such as cost reduction, ease of access to knowledge and access to specialist or complementary skills, for the formation and existence of networks. Coombs *et al.* (1996, Chapter 1) observe that the economic perspective on networks has mainly focused on rational explanations of relationships between firms. In the sociological approach, the focus has been on links and interactions between persons as actors within and between organisations and the processes involved in network relationships. In principle, the economic perspective can be applied to networks of individuals and units within firms and some of the concepts and theories, such as on information and trust, are equally relevant for explaining internal and external personal networking.

Knowledge and Innovation

This section summarises the features of knowledge and innovation which have a bearing on the contribution of external relationships to innovation within firms. While the focus in this section is on technology, similar conclusions can be drawn on the nature of other types of knowledge and capabilities and non-technological innovations. Some technological knowledge is widely available in publications and even in marketed products, but a great deal of such knowledge used by firms is proprietary in nature (Dosi, 1988; Nelson, 1992a). The proprietary knowledge is often highly specific and differentiated and developed in the context of the products and processes of a firm.

The capability to innovate is based on accumulation of knowledge over time. However, it is possible for imitators to acquire technological competence. The ease of imitation depends on the complexity of the technology, the extent to which it is explicit, and the existing knowledge base and capabilities of the appropriator (Mansfield, 1985 and Archibugi and Michie, 1998). Liebeskind (1996) gives the example of Japanese imitation of Western clock making technology in the eighteenth century, by simply dismantling the clocks to observe the workings, as an early example of reverse engineering with no need for a network relationship with Western clock makers.

Competitive advantage depends on the effectiveness with which a firm may be able to protect its innovations and, more crucially, on its capability to innovate. Schendel (1996) refers to "the process by which knowledge is created and utilised in organisations" as the "key inimitable resource" and an important source of economic rents. Firms can protect their technologies by legal and other methods. However, such protection is not complete. Liebeskind (1996) notes that property rights in knowledge are narrowly defined and costly to write and enforce. While, in principle, it is possible to impose legal restrictions on the use of knowledge by others, such restraints are difficult to enforce in practice as technology and other forms of knowledge can be used with modifications to circumvent the restrictions. Further, protection through a patent may be unsatisfactory as it can reveal to rivals precisely the knowledge that the owner seeks to protect.

While technological capability is leaky to a greater or lesser extent, it is not necessarily easily transferable. In general, only part of knowledge is codifiable in blueprints, patents and scientific text. There is other knowledge which is tacit and therefore personal and context dependent. It can be acquired only by long processes of learning (Lundvall, 1996). Therefore knowledge is specific to particular agents such as individuals and firms and its transfer may entail significant costs. Again there are differences between technologies in the ease of transferability. In the case of the clock making technology, it appears to have been relatively easy. In specialised technologies such as electronic controls for advanced machine tools, it is much more difficult. Partly as a direct implication of the above observations and partly as a consequence of the judgement of the agents, the evolution of knowledge is path dependent. In other words, it is highly influenced by the knowledge already accumulated by economic agents (Pavitt, 1988; Dosi, 1988).

Innovations are undertaken in the context of existing knowledge but with a need to learn from external sources. Acquisition of tacit knowledge and capabilities may require a continuing relationship over time with the suppliers who typically look for some reciprocal advantages. This raises the question of the appropriate forms of relationships for exchanges or sharing of information and knowledge between agents and their contribution to facilitating or hindering innovation.

A problem with the literature is that the term network is ascribed different meanings. This chapter uses the term in two distinct (but related) senses. First, there is the network relationship between two (or more parties) engaged in a collaborative venture (implied in the foregoing discussion). Such a relationship may or may not be within the context of a wider network. The second meaning is a broader one in which formal or informal arrangements enable a group of firms and individuals to acquire information, knowledge and resources and pursue common interests. This is similar to Camagni's (1991) notion of the local 'milieu' as a generator of innovative behaviour.

In a world of free and perfect information, a firm communicates with suppliers and customers in a costless way to conduct buying and selling transactions. In practice, however, there are a number of commercial uncertainties facing an enterprise, especially because of the unpredictability of the consequences of dynamic interaction among independent actors outside the firm (Parker and Stacey, 1994). Camagni (1991:3) defines an innovative milieu as "the complex network, of mainly informal social relationships in a limited geographical area, which enhance the local innovative capability through synergistic and collective learning processes". By enhancing capability through shared knowledge and by speeding up diffusion, the local milieu can reduce the cost of acquiring knowledge and the uncertainty that is intrinsic to technological development and innovative processes. It can enable better understanding and control of the possible outcomes of the firm's decisions, technological information and other firms' strategies. Personal relationships play an important part in shaping the development pattern of local production systems.

More or less formal associations between enterprises and public agencies also have a part to play in carrying out some of the uncertainty reducing functions. In the innovation networks approach, more emphasis is placed on innovation as a collective process associated with the milieu than on the entrepreneurial and innovative activities of enterprises. According to Dunning (1995:462), 'Marshall (1920) was one of the first economists to recognise that the spatial clustering or agglomeration of firms with related interests might yield agglomerative economies and an industrial atmosphere, external to the individual firms, but internal to the cluster.'

In a somewhat different conceptualisation, Porter (1990) recognises the dynamism of local clusters, but also draws attention to the competitive rivalry among local firms which enhances competitive advantage. Nelson (1992b) cites Porter (1990) and Freeman (1987) as some of the recent writings in which successful industrial development is attributed to systems involving a mix of private and public institutions. Private institutions include firms, formal networks such as industry associations and professional societies, as well as informal networks. Public institutions include regulatory agencies and enabling agencies which, for example, support research and development, training programmes and extension services.

A network relationship implies the existence of a continuing link between organisations and persons. It may involve economic transactions (sale and purchase of products or services), exchange of information or collaboration in some activities. Even if a formal contract exists, a network relationship typically implies a degree of informal communication, understanding and forbearance. Even the necessarily brief discussion in this section indicates that there are different types of network relationships and motivations for them. These are summarised in Table 1.

For "cost sharing" and "acquiring/sharing information and knowledge" in Table 1, the whole network is the appropriate level of the relationship. The remaining two "acquiring /sharing technology and other knowledge — single transaction" and "acquiring/sharing and developing technology and other knowledge — continuing relationships" are transactions between two or more collaborating actors. The first two include formal associations for exchanging information and sharing costs at a broad level and informal arrangements and contacts through which a great deal of market and technical information and knowledge are transmitted. The last two, and especially the last, represent much closer relationships between partners.

A Transaction Cost Perspective

Reducing the costs and uncertainties associated with market transactions is a possible reason for developing a continuing relationship with suppliers and customers. Therefore transactions cost economics, a model which attempts to explain why some economic transacting goes on within firms or 'hierarchies', while other transactions occur

Table 1. Reasons for networking and forms of network relationships.

	Features	Types of relationships and partners	Comments
Cost sharing	Training, lobbying, industry and R&D information resources.	Industry associations and government agencies.	Main objective is cost saving as cost of internalising would be high. Some activities would be more effective (for example, lobbying) at the industry level than at firm level. Also reduces the free rider problem, for example, where a firm trains employees who move to other firms.
Acquiring/sharing information and knowledge	Information on markets, actions of competitors, information and knowledge on new technologies.	Informal personal and professional contacts. Professional societies, industry associations, suppliers, customers and informal contacts with competitors.	These types of information and knowledge exchanges correspond with the notions of industrial "atmosphere" or "milieu". Communication is mainly on the basis of mutual benefits and in general risks of loss of core knowledge are low, though some information, such as market information, may be highly sensitive. Some capabilities based on existing stock of information and knowledge are necessary to recognise, evaluate and use the information.

Table 1. (*Continued*)

	Features	Types of relationships and partners	Comments
Acquiring/sharing technology and other knowledge - single transaction	One-off acquisition/sale, licensing agreement.	Classical contracts with short term informal exchanges which are not fully specified in the contract.	Requires ability on the part of the acquirer to recognise, evaluate, negotiate, absorb and adapt the technology. A one-off contract could be the beginning of a longer-term relationship.
Acquiring/sharing and developing technology and other knowledge - continuing relationships	R&D collaboration ventures, co-production, sub-contracting, JVs.	Classical contracts with continuing long-term formal or informal relationships.	May complement collaboration in operations and production. Typically implies complementary capabilities and assets, which may be technological, or alternatively one partner may be supplying the knowledge in return for access to market or resources.

within markets, appears to be a logical starting point in developing an economic perspective on networks. The origins of the model go back to Coase (1937) asking why, given the efficiency properties of markets, firms existed at all. Firms were identified by Coase as non-market institutions in which resources are allocated by management fiat. He concluded that, 'a firm will tend to expand until the costs of organizing an extra transaction within the firm become equal to the costs of carrying out the same transactions by means of an exchange in the open market or the costs of organizing in another firm' (Coase, 1937:395). Although Coase did not use the term 'transaction cost' in his 1937 paper, subsequent work by Arrow (1969) and Williamson (1975; 1985) has popularised the term.[3]

Williamson's analysis has become the mainstream within transaction cost economics (for example, Williamson, 1975; 1985; Williamson and Winter, 1993; also see Jensen and Meckling, 1976; Aoki, *et al.*, 1990; and for a critique, Pitelis, 1993). The focal point of the analysis is the costs of arranging, monitoring and enforcing contracts under conditions of imperfect information. Transaction costs are related to *bounded rationality, opportunism* and *asset specificity*. Bounded rationality implies rational decision making by buyers and sellers under incomplete information. Opportunistic transacting, or 'self-interest seeking with guile', is a logical outcome of transacting under incomplete information (Williamson, 1996:6). Given information imperfections, one or other parties to a transaction may be able to exploit their information advantage. An obvious example is the concealing by a seller of unfavourable aspects of the history of a second hand car.

In business relationships, asset specificity[4] creates a problem where one party to a transaction has to invest heavily in tangible assets (for

---

[3]Kenneth Arrow appears to have been the first to use the term transaction costs, in 1969, to describe the 'costs of running the economic system' (Arrow, 1969:84).
[4]Williamson distinguishes six types of asset specificity: site specificity (geographic location of plant and machiery), physical specificity (for example, specialised production technology), human asset specificity; brand name capital; dedicated assets (for example, reserved for a particular company or group of companies), and temporal specificity (time dependency).

example, plant and machinery) or intangible assets (for example, training) to fulfil the contract, implying high sunk costs that restrain market exit (Klein, Crawford and Alchian, 1978). Knowing that one party to a transaction (or a series of related transactions) cannot easily (costlessly) withdraw from transacting, the other party may exploit the situation and use the threat of withdrawal to renegotiate more favourable terms. Economists have termed such circumstances the 'hold-up' problem (Hart and Moore, 1988). Awareness of the threat of 'hold up' explains an unwillingness to engage in a market transaction requiring high sunk costs and therefore, asset specificity is a reason for internalising the transaction.

Dosi (1988) uses the transaction costs perspective to explain why the in-house R&D laboratory is the dominant organisational form for corporate technological research. Market transactions involving research activities may imply (a) incomplete specification of contracts, given the uncertainty about research outcomes, (b) lack of adequate protection of proprietary information and knowledge, (c) possibility of "lock-in" with a research supplier, who can subsequently earn rents from asymmetric advantage, (d) weak incentives for minimising costs and (e) difficulty of monitoring costs (also see Teece, 1988). While vertical integration to supply R&D and other products and services is one possible way of addressing the 'hold-up' problem, it can also impose costs as a result of, for example, lack of competition for supplies and diseconomies of scale in procurement (Grossman and Hart, 1986; Hart and Tirole, 1990).

The transaction costs model would see a network relationship as a relational or implicit contract or series of contracts (Kay, 1993) under which the parties rely on mutual understanding and cooperation to foster commitment, trust and reputation (concepts we return to later) as means of reducing transaction costs. Williamson (1985) recognises that transacting arrangements between the market and hierarchy extremes (for example, repeated trading, subcontracting, franchising and joint ventures) are more common than acknowledged in his previous work and refers to them as 'bilateral governance'.

Although the transaction cost analysis of organisational form is a useful starting point for understanding networks, arguably it is an insufficient explanation and may not even be a necessary one. This follows from some of the weaknesses of transaction cost model, which has been criticised for adopting an over-static view of the world and neglecting the drivers of *change*, including the dynamics of institutional change and notions of cumulative processes in economic development. The model also concentrates upon costs rather than the *benefits* of different governance structures (Dugger, 1983; Pitelis, 1991; Foss, 1993; Dietrich, 1994:37). For example, whether to source internally or externally may be more an issue of resources or capabilities than a matter of transaction costs (Walker and Weber, 1984). Dyer (1997) emphasises the objective of maximising transaction value and not minimising transaction costs in inter-firm collaborations and Ebers (1997) refers to economies of scale and scope, increased responsiveness and flexibility, enhanced learning and capabilities, as motivations for networks.

It is necessary, therefore, to widen the analysis on networks beyond the notion of transaction costs. Certain insights from evolutionary economics are reviewed briefly below to focus on the dynamics of institutional and technological change; and then the resource-based view of the firm is discussed because of its concern with firms' internal capabilities. Both approaches suggest that organisational structures are not the result of a simple deterministic relationship based on transaction costs, but consequences of strategic choices taking account of firms' internal capabilities and change in the external environment.

Evolutionary Economics and Resource-based Theory

Evolutionary economics is concerned with the processes of technological change and innovation. These processes are seen as cumulative and as consisting of potentially diverse development paths.

Evolutionary theories are also compatible with incomplete contracting as a means of dealing with a changing and uncertain environment. Drawing upon Schumpeter's insight that capitalism is essentially an evolutionary process, the role of change is stressed, in which firms grow, decline and perish (Nelson and Winter, 1982; Hodgson, 1993). The emphasis is essentially on disequilibrium positions and change rather than comparative static equilibria and therefore on related concepts of path dependency and time in innovation and technological change (Teece and Pisano, 1994). Evolutionary economics stresses the role of institutions in economies, with institutions defined in terms of norms, routines and rules. This aspect of the analysis links to issues of behaviour, including trust, in economic transacting; an issue returned to later.[5]

In recognising the importance of information imperfections, evolutionary economics complements Williamson's transaction cost economics, but goes well beyond it. Evolutionary economists are generally critical of the comparative static nature of transaction cost theory, and in their view, its underestimation of the impact of information costs in modern economies (Hodgson, 1993). There is much greater emphasis on *changes* in institutions, including firms and markets. There is also recognition of the interaction between economic and social processes within and outside firms, leading to concepts of adaptation and learning. Evolutionary theories emphasise cumulative learning, learning-by-doing and limited search routines, suggesting that proximity to suppliers, customers and competitors may be critical in the innovation process (Zander, 1994).

Nelson and Winter draw on a combination of ideas when developing their analysis of the role of firm-specific routines, including the behaviouralist notions of Cyert and March (1963), Schumpeter's work on entrepreneurship (for example, 1950) and Edith Penrose's seminal treatise on the growth of the firm (Penrose, 1995, ed.). In

---

[5]In this sense, evolutionary economics is much closer than neoclassical economics to the sociological perspective, that economic behaviour is embedded in a set of social relationships, rules and institutional constraints (for example, Granovetter, 1985).

their analysis 'routines' 'are the skills of an organization' (p. 124) and consist of following inarticulate or tacit rules of behaviour. Learning is affected by the social, cultural and economic context in which people operate and markets are identified as but one mechanism for transmitting information and co-ordinating resources (also Nelson, 1996).

The work of Nelson and Winter forms a bridge between evolutionary economics and the resource-based theory of the firm. In neoclassical economics, the nature of the firm is rarely probed and the textbook 'theory of the firm' is simply a theory of price and output determination in different market environments. It has nothing to say about internal resource management. Edith Penrose was one of the first economists to draw attention to internal capabilities, and has since observed: 'Few economists thought it necessary to enquire what happened inside the firm and indeed their "firm" had no "insides" so to speak' (Penrose, 1995:x, Foreword to the third ed.). The transaction cost literature considers internal operations, but as we have seen in terms of transaction costs rather than capabilities. In this sense, it is incomplete. In particular, there is a need to look at how new choice sets evolve and *inter alia* this necessitates attention to the dynamic interaction between organisation and technology (Lazonick, 1993:196–197).

Penrose recognised the importance of organisational learning for improving capabilities of the firms. This idea was further developed by Richardson (1972) exploring the boundaries of the firm, not in terms of transaction costs, but in terms of endowments of resources. Richardson rejected a simple dichotomy between markets and hierarchies (as at that time existed in the transaction cost literature) preferring to view industrial activity as 'a dense network of co-operation by which firms are interrelated'. The stress on a firm's capabilities in both Penrose's and Richardson's work, and its later formulation in 'routines' within firms in the writings of Nelson and Winter and others, has led to a wider recognition of the importance of unique resources and capabilities in explaining competitiveness (Kay, 1993).

Previously, and following the emphasis in neoclassical economics on market structures, generally competitiveness was analysed by strategy specialists in terms of the external environment facing firms, including the degree of competition a firm faced. This is particularly evident in Michael Porter's 'five forces' model (Porter, 1980).[6] Resource-based strategy switches attention to the internal capabilities of firms, including knowledge, skills and experience, as well as command over material and technical resources needed to remain competitive (Langlois, 1997:288). Tangible assets may be easy to emulate, but intangible capabilities, involving routines, specialised skills, culture and collective memories of employees, are not so easy to copy.

This emphasis on knowledge as an intangible asset is now popularised in the notion of 'the learning organisation' (for example, Senge, 1990), where knowledge is considered to be the most important resource, and learning the most important process (Lundvall, 1992).[7] Tacit knowledge is difficult if not impossible to articulate and codify and is embodied in the people who make up the organisation.[8] Such knowledge is communicated through social interaction including

[6]Jarillo (1988:31) suggested that the 'preeminence of models of strategy based on microeconomic theory' has reduced the interest in networks by 'strategy scholars' because 'the construct of networks is difficult to fit within the basic paradigm of competitive strategy'. Following the rise of resource-based strategy, however, it is less problematic to combine competitive and collaborative strategies at the theoretical level. For exampe, Burton (1995) provides a rationale combining Porter's five forces model with a model of collaboration with a view to accessing external capabilities.

[7]In a similar vein, Casson (1998) has suggested that 'information costs' are more pervasive than transaction costs in modern economies. In this context, inter-firm collaboration can be seen as a rational response to the costs of collecting and processing information: 'The pattern of institutions existing at any given time...[is] a rational response to the social need to economise on information costs' (Casson, 1998:153; also see Carter, 1995).

[8]In relation to information there is an obvious overlap here with aspects of Austrian economics. In particular, Hayek (1960:27) stressed the role in market economies of collective learning and experiences passed down over time; the lessons of which are embodied in our institutions, technology, language and ways of doing business.

personal and informal modes of communication (Polanyi, 1966). Sharing tacit knowledge through intensive interaction has been judged the key to knowledge-creation within Japanese industry (Nonaka and Takeuchi, 1995). The resource-based view of the firm is complemented by the 'core competence' literature (Hamel and Prahalad, 1990).[9] From this perspective, the basis for relationships among companies in a network is that each offers a relevant competence that the other lacks (Cravens, Piercy and Shipp, 1998:215).

Since a firm's particular capabilities are determined by resources and past experience and are embodied in technical and organisational expertise or organisational learning and routines, as suggested by Nelson and Winter, this can make radical changes in working difficult or highly costly: 'The choice of new strategies is almost inevitably conditioned by the original resource base of the firm together with the structures and routines that have developed to accommodate past strategies. There is consequently an element of path dependency in the way in which strategy usually evolves' (Jones, 1998:23).

From the discussion of the nature of knowledge in Sec. 2 and evolutionary economics and the resource-based model in this section, it is clear that the development of formal or informal network relation-ships is a means of development of new strategies and capabilities involving tacit knowledge. A formal network relationship can be distinguished from a market transaction in that there may be a longer term contractual agreement to collaborate between firms (for example, in a production venture or technological collaboration). However, even formal agreements require informal understanding and flexibility (see following section). In the context of innovation networks, informal personal network relationships between professionals are important sources of technical and market information. While no monetary transaction takes place in such

---

[9]Similarly, Pettigrew and Whipp (1991) talk about the 'intangible assets' of the firm and Kay (1993) the firm's 'distinctive capabilities'.

information exchanges, von Hippel (1987) found that in "information trading", the supplying person typically assessed (a) the competitive value of the information (it would not be provided if it was crucial for the competitiveness of the supplying person's firm) and (b) the possible value of future reciproca-tion by the recipient of the information. The recipient also recognised the obligation to reciprocate in the future. Araujo (1998) cites the more recent study of the Danish biotechnology industry in which a 'barter economy' of exchange in favours and services exists between firms and research institutes (Kreiner and Schultz, 1993). Powell *et al.* (1996) demonstrate the role of formal and informal networks in accelerating learning in the US biotechnology industry.

A common feature of network relationships is that they are based on an understanding between partners which does not have to be fully enforced by formal contracts. The partners accept certain obligations and reciprocation and understanding if the other party is unable to fulfil its part of the 'bargain'. This requires a level of trust between the parties. Trust that each party will fulfil obligations and not abuse access to information and resources is at the core of a network relationship. We therefore turn to an analytical examination of trust.

The Roles of Reputation and Trust

Recent years have seen a growing interest in the roles of reputation (based on past behaviour) and trust (based on expectations of future behaviour) in business relationships and the working of economies (for example, Kramer and Tyler, 1996; Lane and Bachmann, 1998). It is now increasingly accepted that reputation and trust play an important part in helping to reduce contracting costs and associated uncertainty in business. Fukuyama (1995:7) has claimed that 'a nation's ability to compete is conditioned by a single, pervasive cultural characteristic: the level of trust inherent in a society'. Where innovation is dependent upon networks of social interaction (Freeman, 1994), there is an implied trust relationship based on the sharing of information.

Organisations are embedded within an institutional environment. Based on North's work (1990; 1993; 1994), reputation and trust can be defined as institutions that help to set the 'rules of the game' in any transacting. According to North, institutions include any form of constraint that individuals devise to shape human interaction. They include formal constraints, such as laws, constitutions and rules, and informal constraints, such as norms of behaviour, customs and conventions. Reputation and trust fall within the latter category. Williamson (1993) has recognised the role of trust in reducing transaction costs and therefore determining the comparative costs of governance, but to date trust is not well integrated into transaction cost economics.

Sako (1992; 1998), sees trust as an alternative governance mechanism to markets and hierarchies. In her terminology, *contractual trust* and *competence trust* provide assurance that the supplier can be entrusted to carry out a task to agreed specifications and quality without expensive vetting. By contrast, *goodwill trust* occurs when someone is dependable and can be given discretion to take initiatives and not to take unfair advantage: 'While the roles of contractual and competence trust are specified within existing technical and contractual relationships between trading partners, the role of goodwill trust extends beyond existing relations and includes the transfer of new ideas and new technology. Thus, while contractual and competence trust mainly benefit operational efficiency, goodwill trust also contributes to the dynamic efficiency of productive systems' (Burchell and Wilkinson, 1997:218).

Few economists assume any level of goodwill trust in their models. Trust is understood instrumentally and is couched in terms of rational, calculative action (Casson, 1991). Lyons and Mehta (1997:243) call trust supported by instrumentally rational behaviour, *self-interested trust* (SIT), in which trust is carefully calculated or incentives are created intentionally in direct response to the presence of behavioural risk. A rational actor bestows trust only if calculation suggests that the gain from reciprocated trust outweighs the gain from a betrayal of trust, and when trust relationships are supported by negative sanctions (Coleman, 1990). This contrasts with *socially-oriented trust* (SOT) based on social relations 'experienced in certain normative, or mutually understood, ways…SOT is the product of either an affectual, a traditional or a value-rational behavioural orientation' (Lyons and Mehta, 1997:244, op.cit.).

Sociologists tend to study trust within a SOT-based framework. For example, Zucker (1986) has claimed that there are three ways in which such trust may arise: (1) *process-based trust*, that results from a history of trustworthy interactions; (2) *characteristic-based trust*, arising from identifiable attributes associated with trustworthy behaviour, such as family or religion (clan-based values, as found in parts of East Asia fall into this category); and (3) *institution-based trust*, linked to institutions such as professions and bureaucracies. Process-based trust is tied to past or expected exchange and entails an incremental process of trust building through the accumulation of knowledge (for example, reciprocal, recurring exchange and reputation).[10] Institution-based trust, by contrast, goes beyond particular partners or a transaction and becomes part of 'the external world known in common' (Zucker, 1986:63); that is to say, it is embedded in societal norms and structures and is institutionalised. This can result from traditions, professions and certification and the practices of inter-mediaries, such as banks. It is particularly relevant where trust goes beyond particular group boundaries. Zucker argues that institution-based trust is sufficient to ensure the efficient

---

[10]Process-based trust leads to reputation. Dollinger, Golden and Saxton (1997) find in their experimental study that the better a firm's reputation the more likely it is to be targeted as an alliance partner.

functioning of advanced economies with their complex socio-economic systems.

Lyons and Mehta argue that the distinguishing features of SIT and SOT 'depend crucially on whether we are examining an isolated exchange or part of a sequence of observed exchanges' (op.cit., p. 247). The latter do not have to be between the same parties provided that the outcomes can be observed by parties who may exchange in the future. In game theoretic transacting, approaches to reputation and trust usually build on Axelrod's (1984) 'shadow of the future', in which co-operation emerges when either the players expect repeated games and a lasting relationship or when both players interact intensively with a third person and wish to preserve their reputation. Repeat games, that can overcome 'the prisoners' dilemma', are based on notions of reciprocity; while longer-term relationships are also supportive of the notion of learning that leads to reputation and trust (or higher costs from breaking trust).

Critics of calculative approaches to reputation and trust, such as those inherent in game theory, argue that reputation and trust involve tacit knowledge which is built up incrementally and can change unpredictably (Lane, 1998:5). They also point out that the willingness to make a pre-commitment is situationally and culturally variable (Sako, 1998) and therefore is not easily reducible to a matter of simple calculation. In this context, Luhmann (1979) refers to *system trust*. This is based on the reliable functioning of certain systems. It is maintained by continued affirmative experiences when using the system and is the form of trust associated with confidence in major societal institutions. Lane (1997), in empirical work involving German and British companies, found that there was a higher degree of stability and consistency in systems of social regulation in Germany, as compared to Britain, and this led to more predictable and consensual relations between firms. The resulting greater mutual trust encouraged long-term and closer technological collaboration.

The institutions of reputation and trust, and particularly the notion of system trust, provide a basis for understanding the emergence of networks as trust building entities (Jarillo, 1988; Kreps, 1990; Sydow, 1998:32). Casson has linked transaction cost analysis to the concept of trust to provide insights into the formation of intra- and inter-firm networks (Casson, 1991; 1997; 1998). Transaction cost theory assumes that the rational economic actor sometimes engages in 'immoral' behaviour (acts opportunistically); whereas, 'the network may be characterised as a high-trust co-ordination mechanism linking independent owners' (Casson and Cox, 1997:177). Reputation and trust help the spread of information conducive to innovation by lowering the risk of opportunistic behaviour.

As network members undergo mutual learning and gain experience, uncertainty about actions and reactions of members is reduced. A good reputation is a business asset that reinforces trust within networks and can substitute for contractual controls (Powell, 1990; Gulati, 1998:303). Hence, it is to be expected that the smaller and more stable the network, the more quickly trust building will occur, other things being equal. Other related issues are contract frequency and social connections such as families and clans (Jarillo, 1988:39). Networks that involve frequent contacts between members facilitate learning and networks based on members with common backgrounds may produce speedier trust building. Similarly, the expectation is that in response to greater costs of opportunism, there will be greater incentives to invest in solidifying relations (for example, more personal visits and face-to-face contact). From this perspective, trust is a form of 'social capital' increasing with use.

Reputation and trust may not, however, be sufficient by themselves and may need to be supported by sanctions or penalties for 'deviant' behaviour (Grabher (ed.), 1993). In other words, reputation and trust are supplemented by mechanisms that contain the risk of misplaced trust. The contract is one such mechanism supported by legal redress (Luhmann, 1979:34; Deakin and Wilkinson, 1998); although where there is trust there should be less need for detailed and formal contracting.[11] With SIT a prime purpose of a contract is to use the law to limit opportunism or to achieve co-operation; with SOT there may be no need to specify detailed contingencies or penalties, though contracts may record what has been agreed.

Ring (1997) distinguishes between 'fragile' trust, found in occasional market exchanges and supported by legal contracts, from 'resilient' trust, found in repeated contracting and networking. R&D collaboration may require disclosure of proprietary information pre-contract and parties are therefore exposed to opportunistic abuse. It may be difficult to define R&D outputs in advance so as to include precise details in contracts (Teece, 1998). Classical contracting aims to specify all details of the proposed transaction so as to minimise the opportunity for harmful actions by others, but in terms of innovation activity this may be difficult to achieve. Indeed, it has been suggested that in this environment classical contracting may foster distrust and therefore actually increase the likelihood of opportunistic behaviour (de Laat, 1997). Too much contractual detail might undermine the essential give-and-take that reinforces the social bond that is the essence of trust. For this reason, networks must build 'resilient' trust, which is facilitated by 'credible commitments' rather than legal contracts, *per se*. These commitments may take the form of investment in assets particular to the collaborative project, joint equity holdings, and similar 'sunk costs'. Such actions signal a party's commitment to the collaboration and make exit from the collaboration costly.

---

[11]Maher (1997) undertook detailed interviews in a number of industries and found that automobile manufacturing transactions were governed by bilateral, long-term relationships with buyers and suppliers. Not all of these relationships were governed by written contracts. Arrighetti, Bachmann and Deakin, by contrast, in empirical work involving firms in Germany, Britain and Italy, found that 'relational' contacting, involving higher frequency of contracting as well as asset specificity, may also be associated with a formal agreement and with the use of legal enforceability as a form of security. The role of the legal system in underpinning relational contracting is arguably greater than has been previously recognised (Arrighetti, Bachmann and Deakin, 1997:192).

Spillovers in Innovation

"Acquiring/sharing information and knowledge" as a reason for networking in Table 1 above implies spillover effects or externalities in the form of diffusion of knowledge generated through R&D and innovation. A firm may benefit from the R&D of other firms, while other firms may benefit from its R&D. Such spillovers may arise from formal and informal communication channels and movement of personnel. Where there are appreciable spillovers, innovation becomes a public good and may be subject to under-investment and free rider problems.

In the absence of cooperation, firms invest in R&D independently of other firms and hope to be the first to succeed. This may lead to wasteful duplication of research. Spillovers, free riding and wasteful duplication of effort are standard arguments for co-operation in R&D for companies and for social welfare. Given that network members gain higher spillover benefits than non-members (Kamien, *et al.*, 1992), the success of collaboration depends upon the generation of synergetic effects among members, which in turn leads to welfare enhancing R&D spillovers, and the reduction of 'wasteful' R&D.

However, co-operation in innovation can also be welfare reducing. In particular, competition in R&D may lead to speedier innovation where rivalry induces higher private investment in innovation (Scott, 1993:149). Also, the more firms that are undertaking R&D, the greater the chance, from a social point of view, of success. In this sense, technological rivalry speeds new products to market and this may be lost with co-operation. In this sense, so-called 'wasteful duplication' in innovation may serve a useful purpose in increasing the probability of success (Martin, 1997). Moreover, innovation rivalry is important where it subsequently leads to rivalry in the product market. Co-operation in R&D may raise the likelihood of collusion when new products are launched, leading to welfare loss in the form of higher prices and lower outputs (reduced 'consumer surplus'). Therefore, there is a tension between the arguments for and against co-operation in innovation from an economic perspective, especially with respect to the welfare effects.

Cooperation and Spillovers in Networks: Insights from Game Theory

In discussing reputation and trust (Sec. 5), reference has already been made to the Prisoners' Dilemma (PD) game, a simple one-shot game in which co-operative activity, which would be to the benefit of rational, self-interested agents, is unlikely to occur. It reveals that a strategy based on individual rationality may lead to a sub-optimal joint outcome where each party has incomplete information and lacks trust in the other party (see Kay, 1993, for a simple exposition). A prior agreement to collaborate may not be credible where the dominant strategy is to cheat.

An agreement to collaborate can be credible, however, in a multi-round game ('supergame') of infinite length or where there is no definite end date.[12] Players have to consider the effects of current actions on future payoffs where there is no definite last round. In each round, therefore, the threat to penalise cheating in future rounds is credible. In this case, in principle more or less any outcome may occur depending on the strategies of the players (collaborate, cheat, tit-for-tat or random choices). To take one of the possible outcomes, tit-for-tat, based on the work of Axelrod (1984) the finite repeated game can be resolved by making an altruistic move followed by mimicking the opponent's every subsequent move. The game

---

[12]For a repeated game with a definite end date, backward-induction can be invoked to show why collaboration would not be credible. With non-cooperative behaviour always anticipated in the last period, there is no incentive to be cooperative in the penultimate period; and so on back to the first transaction. In other words, the 'definite end date' game turns out to be equivalent in outcome to the one-shot game.

produces a result consistent with Zucker's (1986) 'process-based trust', which is the product of a history of trustworthy interactions.

In practice, many games in business relationships have a finite length with uncertainty regarding the end date. Even where there is a formal agreement with an end date, there is the possibility of extension or a new agreement at the end. Where there is uncertainty regarding precisely when the game will end, it can be shown that a player has no incentive to cheat if it expects to earn less from cheating than from not cheating. Formally, there will be no incentive to cheat provided:

$$\prod{}^{\text{cheat}} \le (\lambda/\phi)\partial = \prod{}^{\text{coop}}$$

$$0 < \phi \le 1 \tag{1}$$

where $\prod$ is the gain, λ the return into the future given that cheating does not occur, ϕ is the probability of the game being repeated, and ∂ is a discount factor to reflect that future gains are worth less at present value (a gain of ten in year 4, for example, is worth less than a gain of ten in year 0).[13]

This can be further illustrated as follows. Where there is non-cooperation then players (due to a sanction or penalty) may receive a lower pay-off for the remainder of the game. Assume that supergame co-operation leads to a higher pay-off (continuously *a*) compared with a possible one-shot maximum pay-off *b*) where there is non-cooperation, and a lower pay-off in subsequent rounds *c*) due to sanctions and penalties when non-cooperation or a breach of trust occurs. If all players always co-operate (strategy *X*) they receive:

$$X = a + \delta \cdot a + \cdots + \delta^{t-1} \cdot a + \delta^{t} \cdot a + \delta^{t+1} \cdot a + \delta^{t+2} \cdot a + \cdots. \tag{2}$$

But if one player stops co-operating at any point in the game (say round *t*) then all players behave non-cooperatively to it. The player

---

[13]This is identical to the logic that underlies present value (PV) calculations and reflects time preference.

that stops cooperating then receives the following pay-off:

$$Y = a + \delta \cdot a + \cdots + \delta^{t-1} \cdot a + \delta^t \cdot b + \delta^{t+1} \cdot c + \delta^{t+2} \cdot c + \cdots. \qquad (3)$$

For this player it will not be worth deviating from co-operation (adopting defect strategy, Y) provided the following condition holds:

$$X \geq Y \Leftrightarrow X - Y \geq 0 \qquad (4)$$

This means that the present value of the opportunity costs of cheating equals or exceeds the benefit of cheating, i.e., a cooperative strategy outcome is at least as beneficial in terms of pay-off as the defect strategy.

If Eq. 4 generally holds, players benefit from a higher, or at least not a lower, pay-off for co-operation than for non-co-operation. This is determined by the relationship between the pay-offs a, b and c, and the discount rate, δ. The discount factor can be interpreted either as the firm's relative weighting of the pay-off from a future interaction round (time preference) or the probability of the interaction continuing into the next round (the possibility that the game will end).

Game theory provides a rationale for both cheating and cooperation. Rules, contracts and institutions such as trust and reputation that support the success of networks, can be interpreted as means of developing credibility and changing the pay-offs to reduce tension by creating more certainty as to the outcome. If there are either greater incentives to co-operate or penalties for non-co-operation then it is more likely for co-operation to become the dominant strategy. This change could come about, say, through more trust or firms having a more common view of the future, as expressed by δ.[14] Also, where a firm is exiting a network but not closing down then it may well value its reputation for good conduct and therefore will still have an incentive not to cheat. Even where the firm closes, the management

[14]For a discussion of the implications of this for industrial policy and regional clusters, see Elsner, 1998.

may value their reputation in the labour market in which case, again, non-cheating behaviour may dominate.

Players are able to switch cognitively during the process of interaction and revise their position or even change the rules of the game. In other words, networks are likely to be subject to dynamics involving re-evaluation and adaptation that are not easily captured within a well-behaved game. Nevertheless, the study of spillovers and game theory by economists provides important insights into the nature and dynamics of networks, including their limitations in terms of both internalising information and credibility. One particular insight involves the optimal size of networks. The costs of monitoring behaviour can be expected to increase as membership increases, since deviant behaviour by an agent may not be so easily observed. In addition, the present value of not cheating can be expected to be less where gains from collaboration are spread over more players. Such considerations help to explain the probable success of local networks, including regional ones, over national and especially large, international ones. Another way of viewing this is in terms of the discount factor, δ, which can be defined to reflect the probability of a particular actor being encountered again in the future. This will be a function of the size of the grouping and the number of times and ways in which group members interrelate.

Conclusions

From an economic perspective, a network is concerned with the economic property of positive externalities and therefore the structure of interactions amongst agents in the network. Current economic research into networks is exploring interdependencies in the context of increasing returns to scale, path dependency, spillovers and complementarities, and the public good nature of knowledge. The results using game theoretic and other perspectives are very heterogeneous. The research remains largely at the theoretical level at present and further details are beyond the scope of this chapter. This chapter provides a basis, however, from which interested readers can begin to access the literature (papers in Cohendet *et al.*, 1998, are recommended).

The importance of information flows, mutual expectations and resource flows is now recognised in the networks literature (see for example, Ebers, 1997:25; Casson and Cox, 1997:176). But the support for innovation networks from an economic perspective is conditional. In particular, while networks can provide a conducive context for innovation, this is not axiomatic. It is possible to conceive of networks that reduce innovation, for example, where firms with high sunk costs in existing products and processes collaborate to thwart change. Work on lock in and path dependency (linked to evolutionary economics) suggests that vicious circles of failure as well as virtuous circles of success may result. As Conway and Steward (1996:5) point out, 'clusters' can lead to a pooling of ignorance and Kern (1998) found that organisations may become locked in and lose flexibility in industrial districts. In other words, innovation may be better served by encouraging diversity in thinking and operations; while competition is a policing mechanism for promoting cost efficiency that may be lost under networking conditions. Economists worry about the impact of cooperation in networks on competition in the final product market and related welfare implications, suggesting an enhanced role for competition policy when networks proliferate (Khemani and Waverman, 1998).

If partners have little to lose from acting opportunistically for own advantage during the life of the network, then the more at risk the network is of behaviour leading to breakdown. In general, an economic perspective confirms that the smaller the number of organisations in the network and the more stable the network in terms of composition, and the more the members share norms and values, the more likely it is that the network will thrive. This is consistent with the findings of those who have studied the concept from a sociological perspective. It also explains why networks across national boundaries or across industries can be difficult to maintain (Powell, 1996). Frequency of communication promotes trust and communication and may be a function of spatial proximity and task interdependence. Equally, it is to be expected that networks will not be easy to manage successfully where there are low sunk costs leading to fewer constraints on network exit. These aspects deserve a sober appraisal in the context of economic theory and empirical evidence.

The recognition that economic processes and decision making are embedded within social institutions and relationships has stimulated developments in economics and enriched economic analysis. The discussion of different types of trust in the form of SIT and SOT is based on the premise that the former is primarily developed from economic motivation while the latter emerges from social and cultural contexts. This is however too simplistic, as SIT could be a basis for developing a culture of trust. SOT may also have developed, at least partially, from economic rationales. In this sense, it may be better to consider the social and economic processes as being intertwined, instead of the economic being embedded in the social and cultural. Some of the research questions in this area are (a) the relationship between economic motivation and cultural embedding, (b) the process of interweaving of economic and social relationships, (c) the relative importance of economic and social/cultural motivations in networks and especially innovation networks, and (d) the role of competition, market structures and processes in networks and their performance.

References

Aoki, M., Gustafsson, B. and Williamson, O. E. (eds.) (1990) *The Firms as a Nexus of Treaties.* Sage Publications, London.

Araujo, A. (1998) "Knowing and learning as networking," *Management Learning,* **29**(3), 317–336.

Archibugi, D. and Michie, J. (1998) "Trade, growth and technical change: What are the issues?" in *Trade, Growth and Technical Change*, D. Archibugi, and J. Michie (eds.), Cambridge University Press, Cambridge.

Arrighetti, A., Bachmann, R. and Simon, D. (1997) "Contract law, social norms and inter-firm cooperation," *Cambridge Journal of Economics*, **21**(2), 172–195.

Arrow, K. J. (1969) "The organisation of economic activity: Issues pertinent to the choice of market versus non-market allocation," in *The Analysis and Evaluation of Public Expenditure: the PPB System*, Vol. 1 Joint Economic Committee, US Government Printing Office, Washington DC.

Axelrod, R. (1984) *The Evolution of Cooperation.* Basic Books, New York.

Burchell, B. and Wilkinson, F. (1997) "Trust, business relations and the contractual environment," *Cambridge Journal of Economics*, **21**(2), 217–237.

Burton, J. (1995) "Composite strategy: The combination of collaboration and competition," *Journal of General Management*, **21**(1), 1–22.

Camagni, R. (1991) "Local 'milieu', uncertainty and innovation networks: Towards a new dynamic theory of economic space," in *Innovation Networks: Spatial Perspectives*, R. Camagni (ed.), Belhaven Press, London.

Carter, M. J. (1995) "Information and the division of labour: Implications for the firm's choice of organisation," *Economic Journal*, **105**, 385–397.

Casson, M. (1991) *Economics of Business Culture: Game Theory, Transaction Costs and Economic Performance.* Clarendon Press, Oxford.

Casson, M. (1997) *Information and Organization: A New Perspective on the Theory of the Firm.* Clarendon Press, Oxford.

Casson, M. and Cox, (1997) "An economic model of inter-firm networks," in *The Formation of Inter-organizational Networks*, M. Ebers (ed.), Oxford University Press, Oxford.

Casson, M. (1998) "Institutional economics and business history: A way forward?" in *Institutions and the Evolution of Modern Business*, M. Casson and M. B. Rose (eds.), Frank Cass, London.

Coase, R. H. (1937) "The nature of the firm," *Economica*, **4**, 386–405.

Cohendet, P., Llerena, P., Stahn, H. and Umbhauer, G, (eds.) (1998) *The Economics of Networks: Interaction and Behaviours.* Springer-Verlag, Berlin.

Coleman, J. (1990) *Foundations of Social Theory.* Harvard University Press, Cambridge, Mass.

Conway, S. (1995) "Informal boundary spanning communication in the innovation process," *Technology Analysis and Strategic Management,* 7(3), 327–342.

Conway, S. and Steward, F. (1996) "Building networks for cross-border technology diffusion," paper presented at the *Managing Technological Knowledge Transfer Workshop,* Milan, 1–2 February.

Cook K. S. and Emerson R. M. (1978) "Power, equity and commitment in exchange networks," *American Sociological Review,* **43**, 712–739.

Cooke, P. and Morgan, A. (1998) *The Associational Economy: Firms, Regions, and Innovation.* Oxford University Press, Oxford.

Coombs, R., Richards, A., Savioti, P and Walsh, V. (1996) *Technological Collaboration: the Dynamics of Cooperation in Industrial Innovation.* Edward Elgar, Cheltenham.

Cravens, D. W., Piercy, N. F. and Shipp, S. H. (1998) "New organizational forms for competing in highly dynamic environments," *British Journal of Management,* 7(3), 203–218.

Cyert, R. M. and March, J. G. (1963) *A Behavioral Theory of the Firm.* Prentice Hall, Englewood Cliffs, NJ.

Deakin, S. and Wilkinson, F. (1998) "Contract law and the economics of interorganisational trust," in *Trust Within and Between Organisations: Conceptual Issues and Applications,* Lane, C. and Bachmann, R. (eds.), Oxford University Press, Oxford.

de Laat, P. (1997) "Research and development alliances: Ensuring trust by mutual commitment," in *The Formation of Inter-Organizational Networks,* Ebers, M. (ed.), Oxford University Press, Oxford.

Dietrich, M. (1994) *Transaction Cost Economics and Beyond.* Routledge, London.

Dollinger, M. J., Golden, P. A. and Saxton, T. (1997) "The effect of reputation on the decision to joint venture," *Strategic Management Journal*, **18**(2), 127–140.

Dosi, G. (1988) "Sources, procedures and microeconomic effects of innovation," *Journal of Economic Literature*, **36**, 1120–1171.

Drucker, P. F. (1992) *Managing for the Future, The 1990's and Beyond*. Butterworth Heinemann, Oxford.

Dugger, W. M. (1983) "The transaction cost analysis of Oliver E. Williamson: A new synthesis," *Journal of Economic Issues*, **17**(1), 95–114.

Dunning, J. H. (1995) "Reappraising the eclectic paradigm in an age of alliance capitalism," *Journal of International Business Studies*, Third Quarter, 461–491.

Dyer J. H. (1997) "Effective interfirm collaboration: How firms minimise transaction costs and maximise transactions value," *Strategic Management Journal*, **18**(7), 535–556.

Ebers, M. (ed.) (1997) *The Formation of Inter-Organizational Networks*. Oxford University Press, Oxford.

Elsner, W. (1998) "A theory of cooperative industrial policy — model building and practical experience," Discussion Paper No. 22a, Department of Economics, University of Bremen.

Foss, N. J. (1993) "The theory of the firm: Contractual and competence perspectives," *Journal of Economic Perspectives*, **3**, 127–144.

Freeman, C. (1987) *Technology Policy and Economic Performance: Lessons from J.* Pinter, London.

Freeman, C. (1994) "Innovation and growth," in *The Handbook of Industrial Innovation*, M. Dodgson and R. Rothwell (eds.), Edward Elgar, Aldershot.

Freeman, C. (1998) "The economics of technological change," in *Trade, Growth and Technical Change*, D. Archibugi and J. Michie (eds.), Cambridge University Press, Cambridge 16–54.

Fukuyama, F. (1995) *Trust. The Social Virtues and the Creation of Prosperity*. Free Press, New York.

Grabher, G. (ed.) (1993) *The Embedded Firm: On the Socio-Economics of Industrial Networks*. Routledge, London.

Granovetter, M. (1985) "Economic action and social structure: The problem of embeddedness," *American Journal of Sociology*, **91**, 481–510.

Grossman, S. and Hart, O. (1986) "The costs and benefits of ownership: A theory of vertical and lateral integration," *Journal of Political Economy*, **94**, 691–719.

Gulati, R. (1998) "Alliances and networks," *Strategic Management Journal*, **19**, 293–317.

Hagedoorn, J. (1993) "Understanding the rationale of strategic technology partnering: Inter-organisational modes of cooperation and sectoral differences," *Strategic Management Journal*, **14**, 371–385.

Hamel, G. and Prahalad, C. K. (1990) "The core competence of the corporation," *Harvard Business Review*, May–June, 79–91.

Hart, O. and Moore, J. (1988) "Incomplete contracts and renegotiation," *Econometrica*, **56**, 755–786.

Hart, O. and Tirole, J. (1990) "Vertical integration and market foreclosure," *Brookings Papers: Microeconomics 1990*, 205–286.

Hayek, F. A. (1960) *The Constitution of Liberty*. Chicago University Press, Chicago.

Hippel, E. von (1987) "Cooperation between rivals: Informal know-how trading," *Research Policy*, **16**(6), 291–302.

Hislop, D., Newell, S., Scarbrough, H. and Swan, J. (1997) "Innovation and networks: Linking diffusion and implementation," *International Journal of Innovation Management*, **1**(4), 427–448.

Hodgson, G. M. (1993) *Economics and Evolution: Bringing Life Back into Economics*. Polity Press, Cambridge.

Jarillo, J. C. (1988) "On strategic networks," *Strategic Management Journal*, **9**, 31–41.

Jensen, M. C. and Meckling, W. H. (1976) "The theory of the firm: Managerial behaviour, agency costs and ownership structure," *Journal of Financial Economics*, **3**, 305–360.

Jones, S. R. H. (1998) "Transaction costs and the theory of the firm: The scope and limitations of the new institutional approach," in *Institutions and the Evolution of Modern Business*, M. Casson and M. B. Rose (eds.), Frank Cass, London.

Kamien, M. I., Muller, E. and Zang, I. (1992) "Research joint ventures and R&D cartels," *American Economic Review*, **82**, 1293–1306.

Kanter, R. M. (1990) "When giants learn cooperative strategies," *Planning Review*, **18**(1), 15–25.

Kay, J. (1993) *Foundations of Corporate Success*. Oxford University Press, Oxford.

Kern, H. (1998) "Lack of trust, surfeit of trust: Some causes of innovation crisis in German industry," in *Trust Within and Between Organisations: Conceptual Issues and Applications*, Lane, C. and Badmann, R. (eds.), Oxford University Press, Oxford.

Khemani, S. and Waverman, L. (1997) "Strategic alliances: A threat to competition?" in *Competition Policy in the Global Economy: Modalities for Cooperation*, Waverman, L., Comanor, W. S. and Goto, A. (eds.), Routledge, London, 127–151.

Klein, B., Crawford, R. and Alchian, A. (1978) "Vertical integration, appropriate rents and the competitive contracting process," *Journal of Law and Economics*, **21**, 297–326.

Kleinknecht, A. and Reynen, J. O. N. (1992) "Why do firms cooperate in R&D? An empirical study," *Research Policy*, **21**, 347–360.

Kramer, R. M. and Tyler, T. R. (eds.) (1996) *Trust in Organisations*. Sage, London.

Kreiner, K. and Schultz, M. (1993) "Informal collaboration in research and development: The formation of networks across organizations," *Organization Studies*, **14**(2), 189–209.

Kreps, D. M. (1990) "Corporate culture and economic theory," in *Perspectives on Positive Political Economy*, J. E. Alt and K. A. Shepsle (eds.), Cambridge University Press, Cambridge.

Lane, C. and Bachmann, R. (eds.) (1998) *Trust Within and Between Organisations: Conceptual Issues and Applications*. Oxford University Press, Oxford.

Lane, C. (1998) "Introduction: Theories and issues in the study of trust," in *Trust Within and Between Organisations: Conceptual Issues and Applications*, C. Lane and R. Bachmann (eds.), Oxford University Press, Oxford, 1–30.

Lane, C. (1997) "The social regulation of inter-firm relations in Britain and Germany: Market rules, legal norms and technical standards," *Cambridge Journal of Economics*, **21**(2), 197–215.

Langlois, R. N. (1997) "Transaction-cost economics in real time," in *Resources, Firms and Strategies: a Reader in the Resource-Based Perspective*, N. J. Foss (ed.), Oxford UP, Oxford.

Lazonick, W. (1993) *Business Organisation and the Myth of the Market Economy*. Cambridge University Press, Cambridge.

Liebeskind, J. P. (1996) "Knowledge, strategy and the theory of the firm," *Strategic Management Journal*, (Winter Special Issue), **17**(2), 93–107.

Luhmann, N. (1979) *Trust and Power*. Wiley, Chichester.

Lundvall, B.-A. (ed.) (1992) *National Systems of Innovation: Towards a Theory of Innovation and Interactive Learning*. Pinter, London.

Lundvall, B.-A. (1996) "The learning economy: Challenges to economic theory and policy," paper presented at the *Euro-Conference "National Systems of Innovation or the Globalisation of Technology?,"* National Research Centre, Rome.

Lyons, B. and Mehta, J. (1997) "Contracts, opportunism and trust: Self interest and social orientation," *Cambridge Journal of Economics*, **21**(2), 239–57.

Maher, M. (1997) "Transaction cost economics and contractural relations," *Cambridge Journal of Economics*, **21**(2), 147–170.

Mansfield, E. (1985) "How rapidly does new industrial technology leak out?" *Journal of Industrial Economics*, **34**(2), 217–223.

Marshall, A. (1920) *Principles of Economics*. Macmillan, London.

Martin, S. (1997) "Public policies towards cooperation in research and development," in *Competition Policy in the Global Economy: Modalities for Cooperation*, L. Waverman, W. S. Comanor and A. Goto (eds.), Routledge, London, 245–288.

Nelson, R. and Winter, S. (1982) *An Evolutionary Theory of Economic Change*. Harvard University Press, Cambridge, Mass.

Nelson, R. R. (1992a) "What is 'commercial' and what is 'public' about technology, and what should be," in *Technology and the Wealth of Nations*, N. Rosenberg, R. Landau and D. Mowery (eds.), Stanford University Press, Stanford.

Nelson, R. R. (1992b) "Recent writings on competitiveness: Boxing the compass," *California Management Review*, 34 Winter, 127–138.

Nelson, R. R. (ed.) (1993) *National Innovation Systems: A Comparative Analysis*. Oxford University Press, Oxford.

Nelson, R. R. (1996) "Evolutionary theorising about economic change," in *The Handbook of Sociology*, N. Smelser and R. Swedberg (eds.), Sage, New York, 108–136.

Nohria, N. and Eccles, R. G. (eds.) (1992) *Networks and Organizations*. Harvard Business School Press, Boston.

Nonaka, I. and Takeuchi, H. (1995) *The Knowledge-Creating Company*. Oxford University Press, Oxford.

North, D. (1990) *Institutions, Institutional Change, and Economic Performance*. Cambridge University Press, New York.

North, D. (1993) "Institutional and credible commitment," *Journal of Institutional and Theoretical Economics*, **149,** 11–23.

North, D. (1994) "Economic performance through time," *American Economic Review*, **84,** 359–368.

Parker, D. and Stacey, R. (1994) *Chaos, Management and Economics: The Implications of Non-linear Thinking*, Hobart Paper 125, Institute of Economic Affairs, London.

Pavitt, K. (1988) "International patterns of technological accumulation," in *Strategies in Global Competition*, N. Hood and J. E. Vahnle (eds.), Croom Helm, London.

Penrose, E. (1995) *The Theory of the Growth of the Firm*. 3rd Ed., Oxford University Press, Oxford.

Pettigrew, A. M. and Whipp, R. (1991) *Managing Change for Competitive Success*. Blackwell, Oxford.

Pitelis, C. (1991) *Market and Non-Market Hierarchies: Theory of Institutional Failure*. Blackwell, Oxford.

Pitelis, C. (ed.) (1993) *Transaction Costs, Markets and Hierarchies*. Blackwell, Oxford.

Polanyi, M. (1966) *The Tacit Dimension*. Routledge and Kegan Paul, London.

Porter, M. E. (1980) *Competitive Strategy: Techniques for Analyzing Industries and Competitors*. Free Press, New York.

Porter, M. E. (1990) *The Competitive Advantage of Nations*. Free Press, New York.

Powell, W.W. (1990) "Neither market nor hierarchy: Network forms of organisation," *Research in Organizational Behaviour*, **12**, 295–336.

Powell, W.W. (1996) "Trust-based forms of governance," in *Trust in Organisations*, R. M. Kramer and T. R. Tyler (eds.), Sage, London.

Powell, W. W., Koput, K. W. and Smith-Doerr, L. (1996) "Interorganizational collaboration and the locus of innovation networks of learning in biotechnology," *Administrative Science Quarterly*, **41**(1), 116–145.

Richardson, G. B. (1972) "The organisation of industry," *Economic Journal*, **82**, 883–896.

Ring, P. S. (1997) "Processes facilitating reliance on trust in inter-organisational networks," in *The Formation of Inter-organizational Networks*, M. Ebers (ed.), Oxford University Press, Oxford.

Rogers, E. M. (1995) *Diffusion of Innovations*. 4th Ed., Free Press, New York.

Sako, M. (1992) *Prices, Quality and Trust: Inter-firm Relations in Britain and Japan*. Cambridge University Press, Cambridge.

Sako, M. (1998) "Does trust improve business performance," in *Trust Within and Between Organisations: Conceptual Issues and Applications*, C. Lane and R. Bachmann (eds.), Oxford University Press, Oxford.

Schendel, D. (1996) "Knowledge and the firm: Editor's introduction to the 1996 winter special issue," *Strategic Management Journal*, (Winter Special Issue). **17**(2), 1–4.

Schumpeter, J. A. (1950) *Capitalism, Socialism and Democracy*. 3rd Ed., George Allen & Unwin, London.

Senge, P. (1990) *The Fifth Discipline: The Age and Practice of the Learning Organisation*. Century Business, London.

Swan, J. A. and Newell, S. (1995) "The role of professional associations in technology diffusion," *Organization Studies*, **16**(5), 847–874.

Sydow, J. (1998) "Understanding the constitution of interorganizational trust," in *Trust Within and Between Organisations: Conceptual Issues and Applications*, C. Lane and R. Bachmann (eds.), Oxford University Press, Oxford, 31–63.

Teece, D. (1988) "Technological change and the nature of the firm," in *Technical Change and Economic Theory*, G. Dosi, C. Freeman, R. Nelson, G. Silverberg and L. Soete (eds.), Pinter Publishers, London.

Teece, D. and Pisano, G. (1994) "The dynamic capabilities of firms: An introduction," *Industrial and Corporate Change*, **3**, 537–556.

Walker, G. and Weber, D. (1984) "A transaction cost approach to make or buy decisions," *Administrative Science Quarterly*, **29**, 373–391.

Williamson, O. E. (1975) *Markets and Hierarchies: Analysis and Antitrust Implications. A Study in the Economics of Internal Organisation*. Free Press, New York.

Williamson, O. E. (1985) *The Economic Institutions of Capitalism*. Free Press, New York.

Williamson, O. E. (1993) "Calculativeness, trust and economic organisation," *Journal of Law and Economics*, **36**, 453–486.

Williamson, O. E. and Winter, S. G. (eds.) (1993) *The Nature of the Firm: Origins, Evolution and Development.* Oxford University Press, Oxford.

Williamson, O. E. (1996) *The Mechanisms of Governance.* Oxford University Press, Oxford.

Zander, I. (1994) "The tortoise evolution of the multinational corporation: Technological activity in Swedish multinational firms 1890–1990," pub. Doctoral dissertation, Institute of International Business, Stockholm School of Economics, Stockholm.

Zucker, L. (1986) "Production of trust: Institutional sources of economic structure, 1840 to 1920," *Research in Organisational Behaviour,* **8**, 53–111.

Chapter 5

Patterns of Networking in the Innovation Process: A Comparative Study of the UK, Germany and Ireland

James H. Love*

Introduction

Economists have long been interested in the connection between technological progress and economic growth. This in turn has led to an interest in the process by which these are linked and by which new or existing knowledge is commercially applied in the process of innovation.

Within the field of innovation, one of the areas most researched by economists has been the determinants of the pattern and extent of innovation between industries, much of this revolving around the

*The research reported in this paper arose from collaboration with Brian Ashcroft (Fraser of Allander Institute, University of Strathclyde) and Stephen Roper (Northern Ireland Economic Research Centre, Queen's University of Belfast). The author is, however, solely responsible for the views expressed.

Schumpeterian hypotheses that large firm size and/or industrial concentration encourage innovation.[1] This chapter retains insights gained from this literature but attempts to broaden the perspective on the determinants of innovation to allow for factors not normally considered in 'traditional' economic analysis: networking and collaborative activities between firms which may be complements to, or substitutes for, research and development by the firm itself. These factors are considered with reference to the 'new institutional' economics, which recognises that institutional routines and social conventions have important roles to play in shaping economic activity, and gives prominence to, *inter alia*, transaction costs and property rights in the development of such activity.

The insights of the new institutional literature are applied to an examination of the patterns of networking and collaboration in the innovation process among a large sample of British, Irish and German manufacturing plants. These patterns are found to vary substantially between the countries, as is the extent of innovation at plant level. National differences in the internal and external organisation of the innovation process appear to be linked to different institutional structures in the UK and Germany, and to different perceptions of the transaction costs and property rights issues underlying these institutional structures.

An Economics Perspective on Innovation Networks

Underlying much of the conventional economic research in this area is an essentially linear view of the innovation process, in which the promise of higher future rewards encourages firms to invest in the necessary R&D to generate innovations. This gives rise to the concept

---

[1]More recently, the focus of interest in much of the economics literature has been on the use of innovation as a strategic weapon, often couched in terms of game-theoretic analyses of 'patent races'.

of the 'innovation production function', which relates the inputs to the innovation process to the outputs (innovations) in a relatively mechanistic manner. A typical formulation is that of Geroski (1990), who suggests the following relationships for any set of industries indexed by i:

$$\log S_i = \alpha_0 + \alpha_1 \log \pi^e_i + \alpha_2 M_i + \alpha_3 Z_i \tag{1}$$

$$I_i = \beta \log S_i + \mu \tag{2}$$

where S_i is research activity, π^e_i, is the expected post-innovation returns, M_i is the degree of monopoly power in the industry, and Z_i is a vector of other relevant factors. Thus research (usually measured as R&D expenditure per sales) is a function principally of the expected profit from innovation and of the industrial structure of the sector,[2] and innovation in turn depends on R&D.

There are two related assumptions underlying this approach: the first is that innovations come primarily from significant scientific or technological discoveries, and the second is that R&D is both a necessary and sufficient condition for innovation to occur. Both of these assumptions can be challenged. There is evidence not only that the majority of commercially significant innovation is actually incremental in nature, involving the development, application and re-application of existing knowledge with little or no scientific advance (Audretsch, 1995), but also that emphasising the activities of formal research laboratories can substantially underestimate the R&D input and innovativeness of smaller firms (Kleinknecht, 1987), leading to a distorted picture of the relationship between R&D, firm size and innovation.

In addition, the linear view of innovation fails to allow for the possibility that innovation — especially of an incremental kind — may come from sources which have at best an indirect link with any

---

[2]This is a test of the Schumpeterian hypothesis that monopoly power is an aid to innovation.

formal R&D process. Two possible sources are considered here; technology transfer and networking. Evolutionary models of the innovation process suggest the potential importance of technology transfers and networks as sources of new technical knowledge, in addition to an enterprise's own R&D effort (Todtling, 1992). The literature on regional aspects of innovation stresses the possible benefits of technology transfer from parent corporations to branch plants, arising from, *inter alia*, access to R&D facilities operated by the parent or proprietary knowledge developed by the parent, from contacts with external research establishments maintained by the parent, or by being the direct recipients of innovations developed elsewhere in the group (Thwaites, 1978; Oakey *et al.*, 1980). Empirical research suggests this may in part explain the higher innovation rates recorded for multiplant corporations in both the UK and Switzerland compared with their independent indigenous counterparts (Goddard *et al.*, 1986; Brugger and Stuckey, 1987), with the latter study stressing the crucial role of intensive exchange of information within the corporation.

Networks can also be a method of enhancing the potential for innovation without necessarily engaging in R&D. The term 'innovation network' has no universally accepted definition, but the recent review by Freeman (1991) suggests that some degree of consensus is emerging in the use of the term. Networks may be regarded as a form of institutional structure involving a market or quasi-market relationship between firms, the purpose of which is to overcome jointly the uncertainty involved in the innovation process. These are explicit arrangements, which do not include the informal information-sharing arrangements which sometimes exist between firms and their local higher education establishments, or tacit links between firms in a local area.[3] The concern is thus with *formal* innovation networks, but which fall short of full-scale merger or any other form of equity sharing arrangement; specifically, the focus is

[3]These might be more properly identified as *milieu* relationships.

on collaborative or sub-contracting relationships between plants unrelated by ownership (Roper *et al.*, 1996:24–34). Technology transfer is therefore an intra-firm phenomenon while networking involves inter-firm relationships.

Reviewing the empirical research on innovation networks since the pioneering SAPPHO project of the 1960s, Freeman (1991) highlights certain key features of this institutional arrangement. First, "…both empirical and theoretical research has long since demonstrated the importance for successful innovation of both external and internal networks of information and collaboration" (p. 501). Second, the evidence indicates that the use of external networks tends to be just as important for firms engaging in in-house R&D as for those with no R&D capacity. Third, involvement in innovation networking arrangements have increased markedly during the 1980s and 1990s, and have extended to many countries. Finally, networking can be equally effective between firms of roughly equal size, as in the case of Silicon Valley (Saxenian, 1991) or between large and small firms (Lawton Smith *et al.*, 1991).

Three possible approaches to innovation have therefore been identified: R&D, technology transfer and networking. Formal analysis from the economics tradition puts the emphasis firmly on the first of these, but to ignore the possible influence of the others might have the effect of overstating the strength of the link between R&D (and possibly size and monopoly power) and innovation. However, to consider these alternative approaches to innovation does not require us to abandon economic analysis entirely.

Parker and Vaidya (this volume) review other theoretical approaches from an economics perspective, some of which are more amenable to a networks-oriented view of the innovation process. Among these is the transaction-cost perspective, which examines the costs involved in managing internal R&D versus those incurred in contractual research agreements with other parties. This literature emphasises the problems of bargaining and of incomplete contracts in market transacting, with considerable weight being given to the

danger of 'hold-up' which may arise from transaction-specific invest-
ments under conditions of uncertainty (Klein, Crawford and Alchian,
1978; Williamson, 1979). By their very nature, contracts involving
R&D tend to be highly incomplete because of the uncertainty of the
research process, and may be characterised by substantial investment
in both physical and human specific capital. Teece (1988) highlights
the problem of 'lock-in' under these conditions: because of the tacit
knowledge acquired by a contracting party in any external R&D
arrangement, there may be very high transaction costs to incur should
the other party seek to terminate the contract for reasons of under-
performance. What is more, the highly uncertain nature of R&D makes
satisfactory contract completion difficult to define, possibly leading
to a preference for market over hierarchy even where rent-seeking
opportunism is not a major threat. This is supported by the empirical
work of Audretsch *et al.* (1996), who find evidence that the existence
of firm-specific human capital is negatively related to the use of
external R&D.

Parker and Vaidya correctly argue that the simple transaction-
cost perspective cannot by itself provide the full answer to the use
of hierarchy versus market and/or network in any given situation,
first because the transaction cost approach puts all the emphasis on
costs rather then the benefits to be gained from any governance
structure, and secondly because of the relatively static nature of
transaction cost analysis and its failure to take account of the dynamics
of institutional change and development. To this could be added the
excessive preoccupation of much of the transaction-cost literature
with the existence of *opportunism*[4] as a necessary condition for the
existence of transaction costs and so the evolution of non-market or
quasi-market institutional structures. However, some attempts have

---

[4]Opportunism goes beyond mere self-interest and includes an element of 'guile'
(i.e., deceitfulness or cunning). The transaction-cost approach does not assume that
all parties act opportunistically at all times, but that the threat of opportunistic
behaviour may require costs to be incurred by contracting parties in order to avoid
its effects.

been made to preserve the integrity of the transaction cost approach yet adapt to these criticisms. Love (1995; 1997) argues that transaction costs can exist in circumstances where opportunism is not present, as in the case of technology transfer between parties of vastly different technological capability. The problem here can be genuine differences of opinion over contract completion under conditions of extreme uncertainty, giving rise to what Langlois (1992) calls 'dynamic transaction costs', i.e., "the costs of not having the capabilities you need when you need them" (Langlois, 1992:124).[5]

Clearly this is heading more towards a 'capabilities' or resource-based approach to the theory of networks and the firm, which overcomes some of the limitations of transaction cost analysis. But even this more dynamic approach does not tell the full story: we must acknowledge that there is a key role to be played by issues of appropriability, that is the firm's ability to protect and exploit the property rights arising as a result of its research. All firms have reason to fear the possible dissipation of rents which may result from disclosure of R&D findings by a research partner or subcontractor, especially when the research is tacit knowledge embedded in individuals (Teece, 1988). Avoidance of such mechanisms is clearly one method of protection from this problem, but at the cost of failing to take advantage of the cost and risk-sharing advantages of a partner. Kogut (1988) argues that joint ventures will be an appropriate method of engaging in collaborative research, because such mechanisms possess two key advantages over a long-term contractual relationship[6] where there is a high degree of uncertainty over specifying and monitoring performance. First, the joint venture involves mutual commitment of resources (financial and/or personnel), and secondly there are joint ownership and control rights to the outcome of the

[5]The view that opportunism is not necessary to give rise to transaction costs has subsequently become more widely accepted, for example, Conner and Prahalad (1996) and the resulting debate in *Organization Science*.
[6]Kogut sees such a contractual relationship as the most likely alternative to a joint venture in the case of R&D.

research. Together, these two attributes give rise to a 'mutual hostage position' in which neither party has an incentive to shirk on the quantity or quality of their input to the venture (i.e., to act opportunistically) because such action will harm the residual value of the joint venture to the detriment of both parties. Unlike long-term contracts there is no need to specify *ex ante* the precise quantity and quality of inputs at every stage: "Instead, the initial commitments and rules of profit sharing are specified, along with administrative procedures for control and evaluation" (Kogut, 1988:321).

There has been some recent empirical research by economists in Europe which considers the effect of networking and other forms of collaboration on the innovation performance of firms. The importance of this work lies in its retaining the theoretical and statistical rigour of economics, while allowing for the insights on networking to be gained from other approaches. Results are mixed. Arvanitis and Hollenstein (1996) find some positive effect on innovation among Swiss firms which use other firms as external sources of knowledge on which to base innovation, but that this effect tends to be much greater for small firms. They interpret this as a property rights issue, with larger firms more able to generate the appropriate knowledge base from internal sources, and more willing to do so to protect the property rights which develop as a result. By contrast, Brouwer and Kleinknecht (1996) find virtually no influence, either positive or negative, of R&D collaboration or any other form of technology transfer mechanism on the innovativeness of firms in the Netherlands. They also interpret this as an issue of appropriability, arguing that only 'weak' innovators will seek partners because of the need to share the spoils of any resulting collaboration. Contrasting with these findings, Love and Roper (1999) find evidence of considerable use of networking and technology transfer among innovating manufacturing plants in the UK, and that these mechanisms substitute to some extent for R&D effort at the plant level.[7] This study employs a two-stage

---

[7]This study uses UK data from the Product Development Survey described in detail below.

modelling procedure; first, the determinants of R&D, technology transfer and networking intensity are estimated at the plant level, which produces strong evidence that technology transfer and networking are substitutes for, not complements to, R&D in the innovation process. Omitting these important alternatives to research input is therefore likely to over-estimate the effect of R&D on the extent and/or likelihood of innovation. This is important, because the second stage of the econometric estimation indicates that both technology transfer and networking exhibit a positive effect on innovation performance at plant level, independent of the contribution of R&D. This is strongly supportive of the view that the traditional economics view of a linear relationship between R&D and innovation is at best a partial explanation of the whole picture.

Studies such as these suggest that it is possible to retain an economist's perspective on the innovation process and its determinants, yet capture to some extent the subtlety of the networking and technology transfer processes which take place within and between firms. However, they are clearly incomplete for (at least) two reasons. First, these studies concentrate largely on inter-firm or intra-group links in the innovation process, with virtually no emphasis on the nature of network linkages *within* the individual innovating plant or firm.[8] Secondly, few of these studies allow comparative analysis across innovating establishments in different countries; doing so would allow the effects of institutional and policy differences on innovation networks and technology transfer mechanisms to be seen in perspective, and possibly permit conclusions to be drawn which are not available to a single-country study. The remainder of this chapter describes the relevant information gleaned from one of the most comprehensive studies of new product development ever carried out in Europe, involving a large-scale survey of manufacturing plants in the UK, Ireland and Germany, which permits some preliminary conclusions on these comparative issues.

---

[8]One exception to this is Roper (1997) which does examine this issue to some extent in the context of small innovative plants.

The Product Development Survey

The Product Development Survey (PDS), a postal survey of over 15,000 manufacturing establishments in the UK, Ireland and Germany, was conducted between November 1994 and April 1995.[9] The purpose of the study was to discover the extent of product innovation and development at each plant, and develop indicators of how this development was organised both internally and externally. In each country, the sample was structured to allow size-band, regional and industry sector comparisons. Overall response rates of 20.6% in the UK (1722 responses), 25.1% in Germany (1374 responses) and 32.0% in Ireland (533 responses) were achieved. Prior to the analysis survey responses were weighted to allow for sample structuring and differential response rates. Weights for each industry/size-band cell were constructed by comparing sample responses and the 1993 target population of manufacturing firms in each country. The weights were subsequently modified to reflect differences in industrial structure between the UK, Germany and Ireland. Full details of the sampling and weighting processes can be found in Roper *et al.* (1996).

The earlier discussion of the innovation process has implications for how innovation outputs are defined and measured. Measures such as R&D expenditure or employment have the double disavantage of relying on the linear, technology-driven view of innovation, and of having no necessary link to any tangible innovation output (Mansfield, 1984). Patent counts have the advantage of being a clear output indicator from a technological development process but may or may not result in the commercialisation or any positive economic advantage. In the PDS, the process of innovation is regarded as a *business* activity related to, and affecting firms' competitive position. This reflects recent official thinking in the UK (Department of Trade and Industry, 1992), and research from the United States which suggests that almost 90% of commercially significant innovations

---

[9]The survey was funded by a grant under the EU's KONVER initiative.

in manufacturing industry are modest improvements designed to update existing products (Audretsch, 1995). The selected measure of innovation outputs must therefore allow for aspects of innovation which are not necessarily tied to R&D, should reflect the commercial importance of adaptation, improvement and rapid imitation, and should be a direct output-based measure of the *extent* of innovation. Innovation is therefore defined as the number of new or improved products introduced at plant level over the period 1991 to 1993.

Table 1 shows the extent of innovation by country and plant size-band. German plants display a higher proportion of innovating firms in most size-bands, a tendency noted in previous studies of the same countries (Roper and Hofmann, 1993), and which has been interpreted as being in part a reflection of German firms' greater commitment to the process of R&D and innovation (Roper, 1997).[10] As with numerous previous studies of innovation (for example, Kamien and Schwartz, 1982; Acs and Audretsch, 1988; Love *et al.*, 1996) the PDS suggests that the proportion of innovators increases monotonically with plant size; it is also clear that the relatively low level of innovation recorded by UK plants results entirely from the poor innovation performance

Table 1. Percentage of enterprises introducing new or improved products by plant size-band.

	n	20–99	100–499	500 plus	Total
UK	1133	56.0	82.9	93.2	61.0
Ireland	404	64.5	79.8	88.0	68.0
Germany	1184	66.0	81.4	94.7	71.4

[10]It seems unlikely that these differences can be explained by macro-economic factors. The three economies were at different phases of the economic cycle during the relevant period (1991–1993), with Germany experiencing a sharp downturn in manufacturing output while UK output was almost static and that of Ireland growing steadily. *A priori*, one would expect this to *reduce* the relative innovation performance of German firms.

Table 2. Innovations per employee by plant size-band (percentage).

	20–99	100–499	500 plus	Total
UK	0.66	0.13	0.03	0.47
Ireland	0.30	0.09	0.10	0.25
Germany	0.24	0.18	0.05	0.21

of the 20–99 employee plants, which have a markedly lower incidence of innovation than their Irish and German counterparts. Table 2 shows the number of innovations per employee by establishment size-band. For all countries small plants show a markedly higher level of activity per employee than large plants: overall, establishments with less than 500 employees have an average of 0.33 innovations per employee, 6.6 times higher than that for plants with 500 or more employees (average of 0.05). This is precisely the ratio found by Acs and Audretsch (1988) when considering only the most innovative industries in their US sample: overall, Acs and Audrestch found that 'small' (less than 500 employees) plants had just 43% more innovations per employee than large plants.

The average number of innovations per employee among UK respondents was approximately double that of Irish and German respondents. The data in Table 2 suggest that this is principally a function of the smallest size cohort, where innovations per employee shows the largest national variation. Particularly intriguing here is the very high level of innovations per employee among small British plants (twice that of the other countries), while Table 1 indicated that this cohort was largely responsible for the low proportion of innovators among British establishments. This may suggest that the smallest UK plants have some particular difficulty in producing a new or improved product in the first instance, but that once they overcome this initial hurdle they are relatively innovation intensive. This is not a feature of other countries or size-bands.

The crucial role played R&D in the innovation process within the conventional economics paradigm was highlighted earlier. The PDS provides information both on whether responding plants performed any in-house R&D, and on how this was organised i.e., in a formal R&D department or an a more *ad hoc* basis. The presence of a formal R&D department suggests a more systematic approach to the innovation process, which may be associated with improved R&D performance (Table 3). In considering the pattern of results obtained here, it is important to bear in mind that the PDS was a plant rather than company based survey, and so the pattern of industrial ownership between areas will affect the profile of collaboration opportunities available to individual plants. For example, the higher the degree of external ownership within a region, the greater will be the potential for plants within that region to take advantage of specialised group facilities for R&D or other forms of technology transfer. The potential importance of this is illustrated by the differences in the degree of external control of manufacturing capacity between the regions of the UK. In Scotland, 70% of manufacturing employment is in plants owned by companies located outside the region (Ashcroft and Love, 1993), compared to 45% in Northern Ireland and 50% in Ireland (Murshed *et al.*, 1993). Among respondents

Table 3. R&D indicators by size-band (percentage).

	20–99	100–499	500 plus	Total
Undertaking R&D at plant				
UK	49.7	72.8	78.1	54.0
Ireland	54.4	58.5	63.5	55.3
Germany	43.8	63.5	91.0	52.5
R&D department at plant				
UK	12.7	40.3	64.4	20.1
Ireland	12.3	31.6	33.6	18.0
Germany	27.9	54.5	77.3	41.2

to the PDS, 85% of German plants were single establishment enterprises compared to 52% and 56% respectively in the UK and Ireland.
While this pattern suggests that fewer German plants will be able to
take advantage of group facilities than in the UK or Ireland, its wider
impact on either R&D or the innovation process is less clear *ex ante*.
German single-plant companies might adopt independent strategies,
internalising R&D and other elements of the innovation process:
alternatively, they might substitute other external relations for the
intra-group connections which are more readily available in the UK
and Irish respondent plants.

   Table 3 suggests that, despite these differences in ownership
structure, there was little overall variation in the percentage of plants
undertaking R&D.[11] Two points are of interest, however. First, for
each country and size-band, the proportion of firms undertaking in-
house R&D is less than the proportion of innovating firms (c.f. Table 1),
lending support to the contention that in-house R&D is not a necessary
condition for innovation to occur.[12] Secondly, there are clear national
differences in the extent to which formal R&D departments were
established at plant level, with German plants on average more
than twice as likely than their UK or Irish counterparts to have an
R&D department as opposed to a more informal product development arrangement, with the difference strongest among the 20–99
employment size-band. This pattern of R&D commitment may in
part reflect national variations in public policy support for innovation,
with German policy more geared towards mutual co-operation
between small and medium-sized enterprises rather than the UK
model in which support for basic or pre-competitive support tends to
be biased towards large firms (Roper, 1997; Ashcroft *et al.*, 1995).
However, in conjunction with the evidence below on patterns of

[11]Although UK and Irish firms were respectively twice and three times more likely
than their German counterparts to use group R&D resources.
[12]There is, nevertheless, a correlation coefficient of approximately 0.5 between the
presence of in-house R&D and the likelihood of innovation for plants in each country.

internal and external networking, this may indicate a fundamentally different approach to the innovation and R&D process among British and German firms respectively.

Innovating firms in the PDS sample were asked for details of their external relationships with other firms during a variety of stages of the innovation process (Table 4). Each firm was asked to identify whether it employed intra-group linkages (i.e., technology transfer

Table 4. Collaborative mechanisms in the innovation process (percentage).

	Networking		Technology transfer
	Collaborative	Sub-contract	
UK			
Identifying new products	5.6	0.7	13.2
Prototype development	3.9	3.8	10.2
Final design/development	5.7	1.5	8.7
Production engineering	4.2	0.9	6.8
Developing marketing strategy	5.8	2.0	13.7
Ireland			
Identifying new products	11.4	0.2	13.6
Prototype development	6.2	3.8	12.7
Final design/development	7.5	2.7	9.0
Production engineering	5.2	2.5	9.4
Developing marketing strategy	8.5	1.3	17.6
Germany			
Identifying new products	23.6	1.1	2.8
Prototype development	21.5	2.0	3.2
Final design/development	17.5	0.4	2.8
Production engineering	5.9	0.1	2.7
Developing marketing strategy	15.8	1.3	3.9

linkages), and two forms of networking linkages; collaborative or sub-contract. In the transaction-cost terminology these two forms can be regarded as quasi-hierarchy and market respectively. At all stages of the innovation process there was evidence of intra-group and extra-group linkages among a minority of innovating plants, with the proportion of plants engaging in some form of collaboration varying from 11.9% (production engineering — UK) to 27.5% (identifying new products — Germany). The pattern here again shows clear evidence of national variations. As might be expected because of the much higher incidence of independent firms in the German sample, plants from that country showed a markedly lower tendency to use technology transfer linkages than either UK or Irish plants. More interesting, however, is the pattern of networking contacts across the three countries. German plants were heavily involved in collaborative mechanisms with other firms, especially in the early stages of the innovation process, with a very limited use of sub-contracting relationships. While Irish and British plants also generally showed some preference for quasi-hierarchical collaborative arrangements over sub-contracting, this difference was much less marked than for the German sample; for example, the proportion of UK and Irish plants using collaboration in the early stages of innovation averages around one quarter of the German sample. The Irish and UK sample also showed an absolutely greater tendency to use the market approach.

This may in part reflect the fact that public policy support for R&D and product development in Germany has generally been available only for projects involving inter-firm collaboration (Keck, 1993). However, there is also evidence that the motivation of innovating plants for becoming involved in networking varied between the sample countries, with German plants putting much more emphasis on cost sharing and risk reduction, while the UK and Irish plants put a higher priority on accessing the expertise of other plants and especially to accelerating the innovation process (Fig. 1).

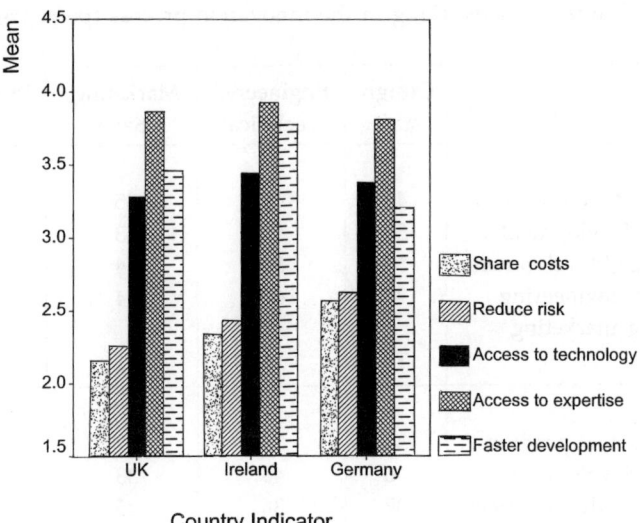

Note: Importance ranked on a scale of 1 (unimportant) to 5 (very important)

Fig. 1. Reasons for networking in product innovation.

It was argued earlier that the present research views innovation principally as a business process rather than a technological process. This suggests that some emphasis should be placed on the organisational aspects of innovation, including internal co-ordination, a topic on which traditional economics models of innovation are largely silent. However, recent models of the innovation process have emphasised the internal network character of innovation, and the importance of feedback between internal functions such as marketing and R&D (for example, Rosenberg, 1982; Bonnett, 1986). Table 5 gives details of the proportion of innovative enterprises involving a variety of functional groups at each key stage of the innovation process. As before, there are substantial national differences in the internal organisation of innovation. Most notable is the much heavier use of design staff throughout the product innovation process by UK and Irish plants than by German plants — indeed, the German sample is notable for

Table 5. Internal networking in the innovation process (percentage).

	Design staff	Engineers/ Technical	Marketing/ Sales	Production staff
UK				
Identifying new products	43	40	85	43
Prototype development	53	52	43	65
Final design/development	55	50	54	63
Production engineering	29	60	14	66
Developing marketing strategy	19	21	91	15
Ireland				
Identifying new products	31	33	82	41
Prototype development	40	43	33	58
Final design/development	37	36	45	54
Production engineering	17	52	12	55
Developing marketing strategy	13	11	88	10
Germany				
Identifying new products	12	62	61	25
Prototype development	20	69	16	52
Final design/development	33	65	34	43
Production engineering	0	57	7	64
Developing marketing strategy	2	17	93	3

the relatively small input of design staff even in aspects of the process which might be considered design-intensive, such as prototype development. Equally marked is the dominance of marketing and sales staff in the initial stages of the product development process among innovating plants in Ireland and the UK: in the earliest stage (identification of new products) over 80% of responding plants from these countries indicated that there was marketing involvement, compared with 61% in Germany.

The relatively high level of involvement of design and marketing staff suggests that British and Irish firms see innovation as a response to market demand and opportunities, which is consistent with their more market-oriented use of collaborative mechanisms highlighted earlier, and with their view of collaboration and/or networking principally as a means of speeding up the process of bringing new products to the market. By contrast, the German emphasis is on technical considerations: Table 5 indicates a much heavier use of engineering and technical staff among German innovators in the first three stages of the innovation process. This too is consistent with the findings on networking discussed earlier, where the use of collaboration to reduce risk and share costs rather than accelerate innovation is consistent with an emphasis on core manufacturing and technological competencies rather than a more market-oriented attitude.

Discussion: Innovation, Networks and Institutions

Several key points emerge from this description of the PDS. First, R&D is important, but by no means necessary, for innovation at the plant level. Other mechanisms, such as technology transfer from other group enterprises or collaborative/networking relationships with other enterprises can assume the role traditionally seen as the reserve of R&D. Secondly, where there is in-house R&D being carried out, among German plants this is much more likely to be organised as a formal R&D department than is the case among respondents from the other countries. Thirdly, there is evidence that patterns of networking vary substantially between plants from the three countries: German innovating manufacturers are more likely to make use of explicit collaborative mechanisms, while those in Ireland and the UK both make more use of a market-oriented subcontracting approach and are more frequent users of technology transfer mechanisms. Fourthly, while UK and Irish firms stress access to others' expertise and acceleration of the innovation process as the principal reasons for networking relationships, German plants put much more emphasis on the more technical and cost-oriented aspects of collaboration, such as risk-sharing. Finally, British and Irish innovators' internal organisation of the innovation process differs markedly from that of their German counterparts, with much greater emphasis on marketing and design input in the earlier stages of the process, while German plants place more stress on engineering and technical staff inputs.

It is possible to relate these patterns of innovation organisation to the theoretical developments outlined earlier. In particular, the differential use of technology transfer and networking arrangements among German, British and Irish firms suggests a difference in attitudes to the potential problems of property rights and transaction costs which may be experienced during innovation. The German emphasis on cost and risk-sharing, and their high use of collaboration suggests they see such mechanisms as having relatively low dangers in terms of dissipation of property rights; they are prepared to accept the lower levels of control which collaboration may give compared with other mechanisms in return for the higher perceived rewards. The knowledge generated by R&D activity is inherently 'leaky', and there is always the danger that firms which have not contributed to the research may free-ride on its benefits. One method of controlling free-rider problems which can arise in such situations is to involve potential competitors which could benefit from the research and development work being undertaken, and the German emphasis on the technical rather than market aspects of the innovation process indicates that this may be another reason for their preference for inter-firm collaboration. British and Irish firms, on the other hand, display a more market-oriented approach which emphasises acceleration of the innovation process and accessing other firms' expertise. This results in less use of collaboration, and more use of mechanisms where the degree of control over technology and the resulting property rights is much higher: technology transfer attenuates the danger of dissipation and free-riding by keeping the technology and tacit knowledge within the group, while subcontracting relationships are useful for maintaining tight control over the use of knowledge and technology because of the clearly delineated nature of both parties' involvement in the contracting process.

There is evidence from other research that these patterns and attitudes run deep within the institutional framework of German and British manufacturing. In an examination of the social regulation of inter-firm relations in the two countries, Lane (1997) finds that although German systems of rule-setting and regulation are highly formalised, this does not supersede more informal relationships. German manufacturing companies develop longer-term and closer relationships with their suppliers and customers than their British counterparts, which in turn encourages technological collaboration. By contrast, the British system of relations between firms does not encourage such behaviour:

> "The establishment and maintenance of effective supplier relations entails higher transaction costs for the firms engaged at every stage of the relationship. The absence of reliable mechanisms of risk reduction makes British managers view long-term commitments with greater wariness than their German counterparts. Close relations of technical collaboration, based on mutual trust, seem to be regarded as less feasible in the British social context." (p. 214)

This work, and the findings of the PDS, suggest a somewhat different perspective on transaction costs and property rights from that to be found in some of the empirical research reviewed earlier. For example, Brouwer and Kleinknecht (1996) argue that collaboration is to be found mainly among weak innovators, because these are the firms which are obliged to share the results of research with others, while Arvanitis and Hollenstein (1996) suggest that collaboration is principally a small firm phenomenon, because such firms lack the internal resources to engage in research and the resulting innovation on their own. Neither of these hypotheses is supported by the PDS data. The prevalence of innovation among German plants of all sizes and their extensive use of collaborative networks suggests that it is unlikely that only the weakly innovative employ this mechanism. And it can easily be shown that collaboration is not the preserve of the smaller enterprise. Table 6 shows the extent of collaboration in five phases of the innovation process among British and German plants both for the sample as a whole and for establishments with fewer than 100 employees. For both countries there is no evidence of greater collaboration among small plants, with the exception of production engineering in both cases. The different national patterns are unlikely to be due simply to different incidences of 'weak' innovators, of small plants, or of other firm-level or plant-level factors; they are embedded firmly in the institutional framework of the countries of which they are part.

Table 6. Collaborative mechanisms by firm size.

	Collaboration (%)	
	All plants	<100 employees
UK		
Identifying new products	5.6	5.2
Prototype development	3.9	3.5
Final design/development	5.7	5.2
Production engineering	4.2	6.1
Developing marketing strategy	5.8	4.5
Germany		
Identifying new products	23.6	19.9
Prototype development	21.5	17.6
Final design/development	17.5	15.9
Production engineering	5.9	7.8
Developing marketing strategy	15.8	13.4

Conclusions

This study has drawn heavily on the Product Development Survey, a postal survey of manufacturing plants in the UK, Germany and Ireland. Although relatively large scale, there are clearly limitations to research of this kind. Although the present research has attempted to examine both internal and external networks, the stress has been very heavily on formal relationships. No light can be shed from research of this kind on the role of the informal links which Conway (1995) identifies as being important for the innovation process. And it is impossible to say at this stage whether the network patterns discussed above are responsible for the different innovation output performance of firms in the three countries, although there are *a priori* grounds for believing this may be the case. Love and Roper (1999) show the importance of technology transfer and networking as a determinant of innovation for the UK firms in the sample, and given the markedly different manner in which German firms organise their innovation inputs it would be surprising if this did not have some effect on innovation performance: work is ongoing which should shed some light on this issue. For example, Lane (1997) finds that the technological collaboration fostered by close inter-firm relationships in Germany may have assisted in innovation development, at least of an incremental, non-radical nature.

Despite these limitations it is nevertheless possible to draw three conclusions. First, the traditional economics perspective does overestimate the role of R&D in the innovation process: other institutional arrangements can substitute for R&D in the innovation process. Secondly, it is possible to retain an economics perspective while examining at least some aspects of the role of networks in innovation: but this involves adopting a broader view of what constitutes the innovation process than is normally permitted in the 'innovation production function' perspective. Finally, and more tentatively, there are national differences in the internal and external organisation of the innovation process which appear to be linked to different institutional structures in the UK and Germany, and to different perceptions of the transaction costs and property rights issues underlying these institutional structures.

References

Acs, Z. J. and Audretsch, D. B. (1988) "Innovation in large and small firms: An empirical analysis," *American Economic Review*, **78**, 678–690.

Arvanitis, S. and Hollenstein, H. (1996) "Industrial innovation in Switzerland: A model-based analysis with survey data," in *Determinants of Innovation the Message from New Indicators*, A. Kleinknecht (ed.), Macmillan, London and Basingstoke.

Ashcroft, B., Dunlop, S. and Love, J. H. (1995) "UK innovation policy: A critique," *Regional Studies*, **29**, 307–311.

Ashcroft, B. and Love, J. H. (1993) *Takeovers, Mergers and the Regional Economy.* Edinburgh University Press, Edinburgh.

Audretsch, D. B. (1995) *Innovation and Industry Evolution.* MIT Press, Cambridge, Mass.

Audretsch, D. B., Menkveld, A. J. and Thurik, A. R. (1996) "The decision between internal and external R&D," *Journal of Institutional and Theoretical Economics*, **152**, 517–530.

Bonnett, D. (1986) "Nature of the R&D/marketing co-operation in the design of technologically advanced new industrial products," *R&D Management*, **16**, 117–126.

Brouwer, E. and Kleinknecht, A. (1996) "Determinants of innovation: A microeconometric analysis of three alternative innovation output indicators," in *Determinants of Innovation the Message from New Indicators*, A. Kleinknecht (ed.), Macmillan, London and Basingstoke.

Brugger, E. A. and Stuckey, B. (1987) "Regional economic structure and innovative behaviour in Switzerland," *Regional Studies*, **21**, 241–254.

Conner, K. R. and Prahalad, C. K. (1996) "A resource-based theory of the firm: Knowledge versus opportunism," *Organization Science*, **7**, 477–501.

Conway, S. (1995) "Informal boundary-spanning communications in the innovation process: An empirical study," *Technology Analysis & Strategic Management*, **7**, 327–342.

Department of Trade and Industry (1992) *Innovation, Technology and Change: Help for Business*. HMSO, London.

Freeman, C. (1991) "Networks of innovators: A synthesis of research issues," *Research Policy*, **20**, 499–514.

Geroski, P. (1990) "Innovation, technological opportunity, and market structure," *Oxford Economic Papers*, **42**, 586–602.

Goddard, J., Thwaites, A. T. and Gibbs, D. (1986) "The regional dimension to technological change in Great Britain," in *Technological Change, Industrial Restructuring and Regional Development*, A. Amin and J. B. Goddard (eds.), Allen and Unwin, London.

Kamien, M. and Schwartz, N. (1982) *Market Structure and Innovation*. Cambridge University Press, Cambridge.

Keck, O. (1993) "The national system for technical innovation in Germany," in *National Innovation Systems*, R. Nelson (ed.), Oxford University Press.

Klein, B., Crawford, A. and Alchian, A. (1978) "Vertical integration, appropriable rents and the competitive contracting process," *Journal of Law and Economics*, **21**, 297–326.

Kleinknecht, A. (1987) "Measuring R&D in small firms: How much are we missing?" *Journal of Industrial Economics*, **34**, 253–256.

Kogut, B. (1988) "Joint ventures: Theoretical and empirical perspectives," *Strategic Management Journal*, **9**, 319–332.

Lane, C. (1997) "The social regulation of inter-firm relations in Britain and Germany: Market rules, legal norms and technical standards," *Cambridge Journal of Economics*, **21**, 197–215.

Langlois, R. N. (1992) "Transaction-cost economics in real time," *Industrial and Corporate Change*, **1**, 99–127.

Lawton Smith, H., Dickson, K. and Lloyd Smith, S. (1991) "'There are two sides to every story': Innovation and collaboration within networks of large and small firms," *Research Policy*, **20**, 457–468.

Love, J. H. (1995) "Knowledge, market failure and the multinational corporation: A theoretical note," *Journal of International Business Studies,* **26,** 399–407.

Love, J. H. (1997) "The transaction cost theory of the (multinational) firm: A note," *Journal of Institutional and Theoretical Economics,* **153,** 674–681.

Love, J. H., Ashcroft, B. and Dunlop, S. (1996) "Corporate structure, ownership and the likelihood of innovation," *Applied Economics,* **28,** 737–746.

Love, J. H. and Roper, S. (1999) "The determinants of Innovation: R&D, technology transfer and networking effects," *Review of Industrial Organization,* **15,** 43–64.

Mansfield, E. (1984) "Comment on using linked patent and R&D data to measure inter-industry technology flows," in *R&D, Patents and Productivity,* Z. Griliches (ed.), University of Chicago Press, Chicago.

Murshed, S. M. *et al.* (1993) "Growth and development in the two economies of Ireland: An overview," presented at *Growth and Development in the Two Economies of Ireland Conference,* Belfast.

Oakey, R. P., Thwaites, A. T. and Nash, P. A. (1980) "The regional distribution of innovative manufacturing establishments in Britain," *Regional Studies,* **14,** 235–253.

Parker, D. and Vaidya, K. (1999) "An economic perspective on innovation networks," mimeo, Aston Business School, Birmingham.

Roper, S. (1997) "Product innovation and small business growth: A comparison of the strategies of German, UK and Irish companies," *Small Business Economics,* **9,** 523–537.

Roper, S., Ashcroft, B., Love, J. H., Dunlop, S., Hofmann, H. and Vogler-Ludwig, K. (1996) *Product Innovation and Development in UK, German and Irish Manufacturing.* Northern Ireland Economic Research Centre/Fraser of Allander Institute, University of Strathclyde.

Roper, S. and Hofmann, H. (1993) *Skills, Training and Company Competitiveness.* Northern Ireland Economic Research Centre, Belfast.

Rosenberg, N. (1982) *Inside the Black Box: Technology and Economics.* Cambridge University Press.

Saxenian, A. L. (1991) "The origins and dynamics of production networks in Silicon Valley," *Research Policy*, **20**, 423–437.

Teece, D. (1988) "Technological change and the nature of the firm," in *Technical Change and Economic Theory*, G. Dosi *et al.* (eds.), Pinter, London.

Thwaites, A. T. (1978) "Technological change, mobile plants and regional development," *Regional Studies*, **12**, 445–461.

Todtling, F. (1992) "Technological change at the regional level: The role of location, firm structure and strategy," *Environment and Planning A*, **24**, 1565–1584.

Williamson, O. E. (1979) "Transaction cost economics: The governance of contractual relations" *Journal of Law and Economics*, **22**, 3–61.

Chapter 6

Shaping Technological Trajectories Through Innovation Networks and Risk Networks: Investigating the Food Sector

Fred Steward

Innovation and Risk in the Food Sector

Technological innovation in the food sector is characterised by significant environmental impacts and a highly sensitive consumer context. The implications of genetically modified organisms in agriculture and concerns over the biosafety of new biological or chemical agents in the food chain are cases attracting recent concern. The food sector spans a complex range of activities, from primary gathering in nature such as sea fishing, through animal and crop based agriculture to advanced food manufacturing industries. Innovation develops in a variety of institutional contexts and draws on a wide range of R&D activities ranging from agricultural science to biotechnology and includes both private and public sector research organisations. Risks to human health and the natural environment are of concern to a visible and diverse array of social movements concerned with countryside preservation, animal welfare, and consumer safety (Tansey and Worsley, 1995).

The institutional diversity of the sector and the breadth of environmental and consumer concerns pose a particular challenge for the social analysis of the food innovation process. Research needs to engage effectively with a large variety of public and private actors in the domains of business, government, knowledge and civil society. Because of the range of the sector there are many specific subdomains dealing with different parts of the industry (fishing, agriculture, food processing etc.) and with different dimensions of policy (economic, environmental, health etc.). This is further complicated by the presence of strong national traditions which in the European Union are undergoing a process of harmonisation.

An earlier study (Steward, 1995) analysed food innovation in terms of competition between different technological trajectories over periods of several decades influenced by the organisation and balance of business, government and public interests. A current study (Steward *et al.*, 1998) seeks to develop this further by drawing on social network analysis in order to explore the relationship between the innovation process and controversies over risk. It aims to create a knowledge base of value to the development of European environmental policy as expressed, for example, in the European Commission's programme 'Towards Sustainability (European Commission, 1993) with its commitment to: 'The central importance of the consumer in reorienting technological priorities towards sustainability; the need to integrate different policy dimensions in order to develop effective sectoral strategies for sustainability; and the creation of structures facilitating the participation of citizens and non governmental actors'.

It has been recognised for some time that the promotion/ stimulation of new technologies and the regulation/control of harmful environmental consequences tend to occur as two separate social processes: a 'bifurcation of these twin dimensions of modern technology' (Irwin and Vergragt, 1989). They criticise the 'separation between the domains of innovation and regulation' in academic studies by contrasting those which focus on the private firm, the economic domain and technological performance with those which concentrate on the public domain, policy processes and technological risks.

They call instead for innovation and regulation to be viewed as part of the 'same social and technical process' and for research to focus on the social and institutional interactions and negotiations of the 'groups and actors — both within and outside industrial firms — which shape innovation'. This shaping is seen to occur through the establishment of a 'dominant problem definition' from the specific views of the problems to be solved held by the different actors. They begin to define the importance of actors at a finer level than that of the firm or state but while calling for analytical attention and empirical investigation of these interactive processes the case study is the only methodological approach that is explicitly mentioned (Irwin and Vergragt, 1989).

Network Conceptualisations Drawing on Evolutionary Models of Technology

From the early 1990s, there has been a growing number of studies drawing on the evolutionary economics tradition which have utilised network concepts in order to develop a systematic theoretical framework for exploring the relationship between innovation and risk.

Firm Based Networks

Groenewegen and Vergragt (1991) explicitly employ the theoretical framework of social networks to investigate the relationship between the innovation process within the firm and the wider context of public environmental concerns. The formation and expansion of social networks between intrafirm actors (for example, R&D groups, marketing units etc.) and with extrafirm actors (for example, customers, government agencies etc.) are analysed in relation to the direction of technological change.

Networks are considered to be created through a process of negotiation between different actors about a common 'problem definition' which, once established, remains relatively stable along with the social network structure of the actors who share it. In contrast with the social constructivist approach the formation and development of social networks is constrained by 'preexisting organisational structures' and the authors are interested in how changes in networks may be brought about, through, for example, the insertion of new specialised environmental units.

Three layers of internal and external networks are identified: the production network of actors such as manufacturing sites, suppliers and customers, the innovative network of actors such as research units and marketing groups, and the strategic network of top management and corporate staff. The study of five chemical firms showed that companies generally acted in anticipation of regulation with the internal production and innovation networks adapting toward acceptance, as did the external strategic network in some cases while in others it was employed to resist it. This showed a difference between the role of the 'technical' networks (production and innovation) and the 'policy' network (strategy/external).

The specialised environmental units varied in their network location between firms. In some they were restricted to the production networks while in others they also had influence in the innovation and strategy networks. The study identified different types of organisational positioning of environmental units but the patterns of network relationships were not explored in a systematic fashion. The need for the development of indicators of 'organisational structure' is identified but not pursued.

Schot (1992) locates the innovation/risk problematic in the framework of quasi-evolutionary technology dynamics. This is proposed as an alternative to the independence and separateness of selection and variation in neo-Schumpeterian evolutionary economics or their coincidence and simultaneity as suggested in the coevolutionary 'seamless web' models of actor network sociology. The Schot model emphasises instead the 'linked by actors' relationship between variation and selection.

This linkage may occur in three ways: *'ex ante selection'* by anticipative adjustments of heuristics by firms; 'strategic niche management' which attempts to create a niche to protect desired variation; and thirdly, a 'nexus' of 'institutional links' which 'mediate between technological opportunities and environmental requirements' through 'active efforts' and a 'learning process'. This relational model highlights the role of three different types of actor: internal actors who 'determine the content of variety generation', external actors who try to 'selectively influence' variation, and nexus actors who 'couple variation and selection'. The third role draws explicitly on the relational roles of organisational sociology: translator, gatekeeper and boundary-spanner.

Although acknowledging that such a role may be played by departments or individuals within a firm Schot's discussion is confined to the three institutions at the formal organisational level: marketing, environmental and quality assurance departments. The importance of interaction between actors is emphasised by a suggested new role for public authorities: 'Creating networks between actors and establishing and enforcing the rules of the game for these networks'. The analysis places networks centre stage but does not draw on or develop social network theory in relation to them.

Two Types of Social Network

Coombs (1995) argues that the emergent synthesis of evolutionary economics and actor-network sociology provides an intellectual framework for transcending the traditional promotion/control dichotomy and involves a 'network-based understanding' of the innovation process. The approach suggests that 'interest...are not... defined by some over-arching social order, but are in fact continuously redefined and reconstructed, in parallel with the creation and use of technology itself' and that 'large scale features of the political and economic structure…are made present in concrete micro-situations not as fixed and unambiguous constraints but rather through the interpretations and discourses to be found amongst the actors in a particular network'. He argues that 'the careful analysis of networks and their role on constituting technologies in society, is a potentially important contributor to the evolution and outcomes of such networks' (Coombs, 1995).

The 'interface between the variety generation process and the selection environment' is highlighted with particular attention to the internal selection environment within the firm [where]…networks… act as vectors to concentrate the effects of broader social networks on the emerging technologies and their properties. An understanding of 'the networks which create technology in the firm' will enable an exploration of 'the connections between two categories of social networks…networks which articulate the perceived contributions of technologies and products to the profitability and survival of business firms, and…networks which articulate the evaluations of technologies and products made by citizens, consumer groups, and public bodies of various types' (Coombs, 1995).

Coombs own exploration of the challenging research agenda posed by this conceptualisation of two categories of social network is limited in the one case to a consideration of the formal location of technology and product strategy in large firm R&D organisation and in the other case to a proposal for new public 'product specifying' research institutes. This is motivated by a brave attempt to identify practical possibilities for policy interventions to 'amplify the connections' between the two categories of network and which are based on current knowledge and avoid variety reduction.

While it suggests intriguing parallels between the new organisation theory approach to the emergent, processual and political nature of strategy and actor network theory it does not pursue the implications of the framework for a more open empirical investigation of actually existing social networks and their interface.

Network Approaches to Organisation and Policy Processes

The emerging paradigm drawing on evolutionary economics which focuses on network relationships between variation and selection appears a fruitful one which raises many intriguing questions for further research. One of these is the degree to which the two types of network can be regarded as analytically or substantively similar, or whether they are distinct as suggested by Blauwhof (1994) who contrasts micro variation processes where the 'economic actor' is the referent and selection processes which have a discourse or communication network as its referent. Another is the relative significance of formal organisational solutions to the interface between networks compared with the need for deeper understanding of the nature and role of informal relationships.

A closer integration of the insights of technology studies with those from organisation and policy studies might offer a productive way forward. Network concepts have been increasingly employed from an organisational perspective to analyse the innovation process in the firm. The policy oriented fields of risk controversies and policy development in network terms. These fields offer some rich resources for elaborating the theoretical framework further as a basis for empirical exploration.

Innovation Networks

Freeman's (1991) review of innovation networks emphasises the pervasive and often informal nature of networks within the firm and their relationships externally. It essentially argues that the empirical tradition of innovation research from the early 1970s points to the importance of networks, even though the term is rarely used explicitly.

Conway and Steward (1996) have reviewed the literature which points to such an approach. Technological innovation should not be viewed as resulting from a single idea, but from a *bundle* or *ensemble* of ideas, information, technology, codified knowledge and know-how, which may or may not be embodied within the new product or process. Furthermore, new ideas seldom appear fully formed and articulated from a single source. This implies that innovation generally arises from a portfolio or network of actors and relationships. Studies of successful technological innovation have highlighted the importance of a number of key characteristics of these 'innovation networks': the key role of external sources and boundary-spanning activity; the diversity of internal and external actors involved in the development process; and the importance of informal or personal relationships in supplementing and 'breathing life' into formally prescribed relationships (the organisation chart) and linkages at the level of the organisation (for example, joint ventures).

Studies have also indicated the importance of managing relation-ships across internal interfaces, such as between project groups, functional departments, and divisions. In particular, research has highlighted the importance of the internal marketing and R&D interface. A diverse range of external sources have been found to contribute to the development of successful innovation, including research organisations, suppliers, competitors, users, consumers and distributors

A number of other studies have indicated the importance of informal or personal *boundary-spanning* contact to the innovation process and particularly in relation to the transfer of *tacit* knowledge (MacDonald and Williams, 1993). The importance of network relation-ships in the innovation process, shown through these studies, is reflected in the growing number of investigations into particular types of these relationships, particularly external inter-firm collaboration. Studies which seek to reveal the relational diversity of the innovation process are far less common (Tidd, 1995). The reliance on the case study methodology in such studies poses problems for comparative work and this has let to interest in new approaches which can capture such diversity in a more systematic analytical fashion.

What is of particular interest is to explore whether certain theoretical propositions about network characteristics and associated behaviour can be tested in a new and more rigorous fashion. One set of such propositions is essentially concerned with the emergence of novelty in any such system and there are a variety of network characteristics such as 'openness' and 'weak ties' which suggest favourable conditions for new ideas to be identified and utilised. Another set of theoretical frameworks is more concerned with how choice is exercised between a variety of different technological possibilities and why a particular option is selected.

The actor network approach of Bruno Latour and Michel Callon (Callon and Latour, 1981) argues that the process of innovation should itself be viewed as the formation of a network of human and non-human actors through a process of enrolment, translation and closure. The emergence of a particular technological path is an expression of the capacity of an individual actor to construct such a stable network. This capacity is seen to rest on translation and enrolment abilities. The power of the actor network approach has been to reconceptualise the process of innovation in a completely relational fashion and refocuses the analysis of innovation toward the reconstruction of networks. These networks are no more or less than the expression of a set of discursive interactions.

A number of authors have located themselves on actor network foundations but have balked at the relativism of the Latourian approach and have sought to retain some notion of structural influences on power and outcomes Elzen *et al.*, (1996) employ the concept of sociotechnical networks which are defined as "those interactions that have some relationship with the development of the artefact" and recast the issue of structural influences as those of 'preexisting networks' on the prospects for the 'development of new configurations of actors'.

Mapping Innovation Networks

Visual mapping of innovation networks offers the possibility of portraying the full pattern of significant relationships in different cases of innovation. The development of any technique for mapping such networks needs to recognise certain key characteristics as of central importance. It must have the power to systematically reveal a diversity of actors — internal and external — and a diversity of relationships — informal and formal. Many forms of depiction do not capture this diversity. At the heart of developing a suitable approach is the fundamental issue of specifying more precisely what is to be included in the 'innovation network' under investigation. The definition of a boundary which is based on a clear concept of the innovation process is critical.

Steward and Conway (1998) have used such an approach in the analysis of environmental innovations in the UK and Germany. Through the mapping of the focal action-sets it was possible to compare and contrast the pattern of the networks mobilised in the development of each of the environmental innovations. Analysis of these graphical representations revealed that the typical morphology of the UK and German focal innovation action-sets were found to differ in two main respects: firstly, the UK action-sets exhibited strong linkages into the research-base, while the German action-sets highlight linkages to the regulatory and public domain; secondly, the UK action-sets frequently incorporated two focal organisational actors (one research-orientated and one market-orientated) to bridge the gap between the research-base and the market-place. Thus, when the action-sets are mapped-out using the segmented templates, the UK cases most often exhibited linkages into the upper-left (knowledge) and lower-right (customer) segments, while the German cases most often revealed linkages into the upper-right (regulatory and political-cultural environment) and lower-left (supplier) segments. The knowledge/customer pattern of innovation network is shown in the UK Hotwork International case (Fig. 1) and the supplier/regulator pattern in the German Brauerei Felesenkeller Helford case (Fig. 2).

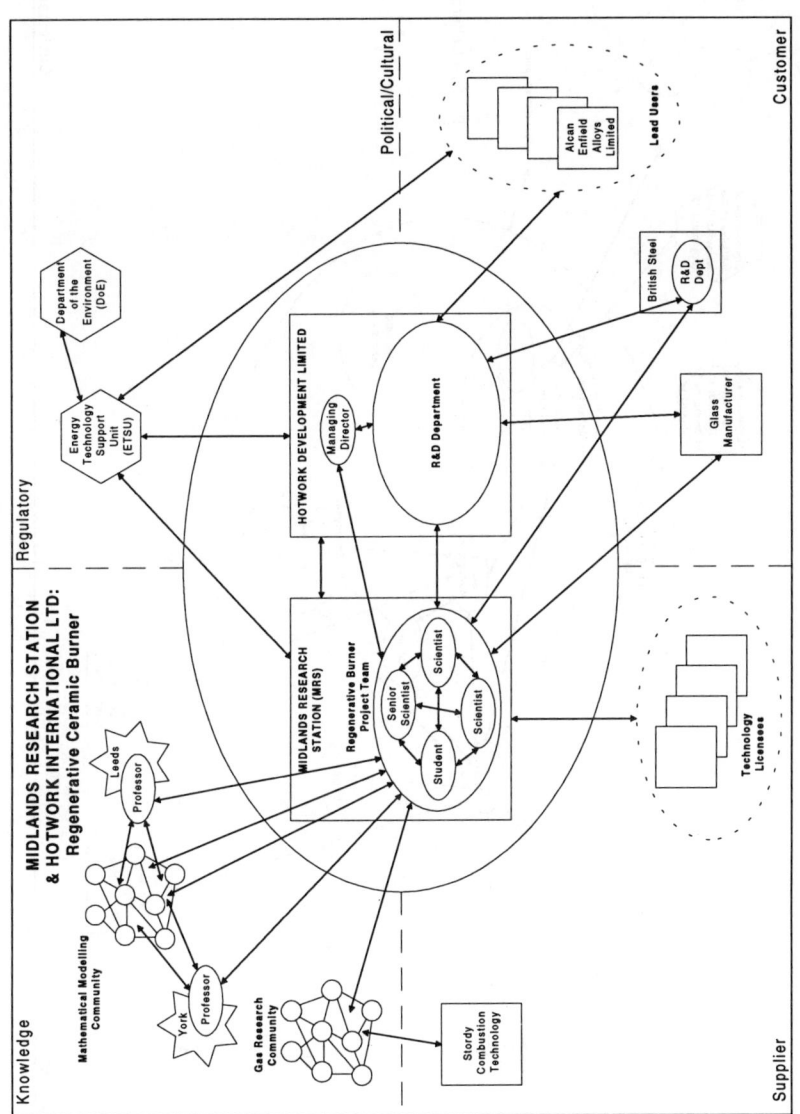

Fig. 1. Actors and links mobilised in the development of the regenerative burner (UK).

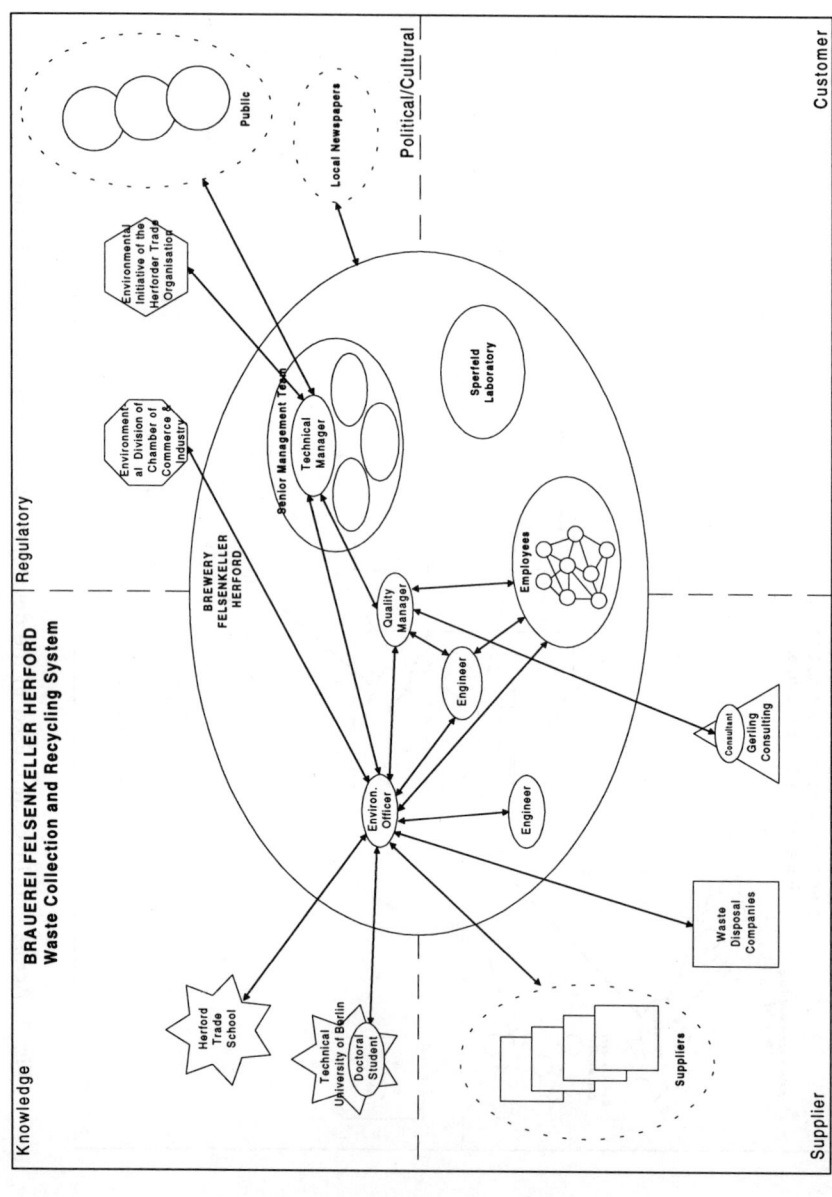

Fig. 2. Actors and links mobilised in the development of a waste collection and recycling system (Germany).

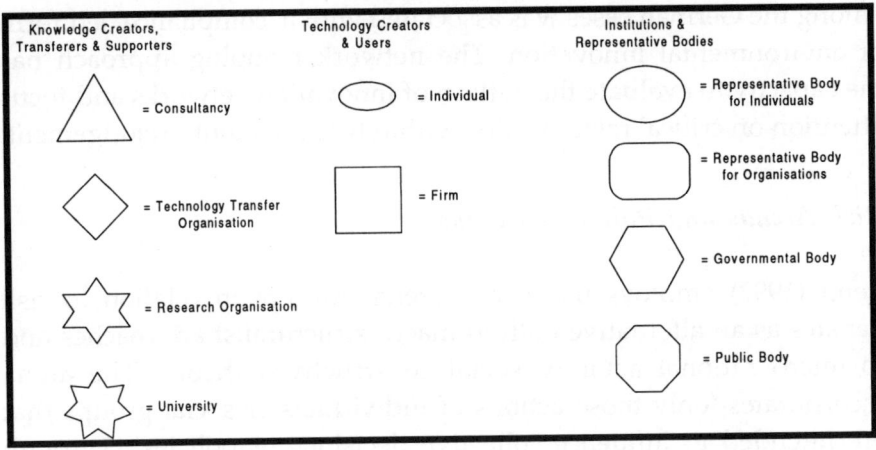

Key to Figs. 1 and 2: Actor types.

The research method adopted in the investigation proved to effective in revealing the diversity of interactions in environmental innovation networks. Both sets of cases show that management of communication across a range of different organisational interfaces is an essential part of the innovation process and requires capabilities for handling multiple discourses. The results show the inadequacy of prescriptions for organisational greening which privilege a particular type of interface or mode of communication.

Two main contrasting patterns of innovation network were found in the study. The knowledge/customer network type with strong linkages into the research-base was commonest in the UK set of cases while the regulator/supplier network type highlighting linkages to the regulatory and public domain was commonest in the German set of cases. The UK networks frequently incorporated inter-organisational collaboration between two focal actors (one research-orientated and one market-orientated) while the German networks often had a more formalised internal environmental management function.

The knowledge/customer type of network prominent among the UK cases was associated with an 'opportunity' led form of environmental innovation. The regulator/supplier type of network prominent among the German cases was associated with a 'compliance' led form of environmental innovation. The network mapping approach has the capacity to evaluate the pattern of innovation networks and focus attention on critical relationships within them without prejudgement.

Risk Arenas and Policy Networks

Renn (1992) employs the 'social arena' concept in relation to risk debates as an alternative both to macro structuralist approaches and to micro rational actor or social constructivist theory. The arena incorporates 'only those actions of individuals or social groups that are intended to influence collective decisions or policies'. Through this it aims to elucidate a set of actors and the 'dynamics of their interactions'. Although the approach seeks to identify the arena rules and the communication patterns between actors 'the fundamental axiom is that resource availability determines the degree of influence for shaping policies'. A focus of the approach is on the ability of actors to mobilise resources.

The arena concept seeks to explain the policy process by analysing the 'symbolic location' of the actors involved. Such symbolic locations are 'neither geographical entities nor organisational systems' and the arena concept therefore avoids prejudging which structural relationships may be of significance. It is defined by the characteristics of a specific autonomous policy field which also includes different institutional contexts (science, law etc.) which Renn terms 'stages'. The arena is characterised by a set of formal and informal 'rules' which set structural limits to the options for action, yet can themselves be modified by novel forms of political action and 'often behave like indeterministic or nonlinear systems' with small changes in strategies or rules able to produce major changes in conflict outcomes.

The range of resources mobilised by actors in the arena range from monetary wealth and legal authority to powers of persuasion, cultural meaning and evidence. 'Communication' between actors in the process of conflict over risk is seen as the mode for 'gaining or exchanging resources' and for 'defining the stakes' of each actor in the arena. Discourses in which actors show a 'capacity for empathy' or 'envision the common good' may lead to agreements beyond a utilitarian balance of individual interests. Risk arenas show characteristics of 'highly ambiguous evidence presented by different actors, weak rule enforcement agencies and a lack of immediate personal experience about the potential consequences of political decisions' which leads Renn to conclude that 'the generation and distribution of resources relies almost entirely on the success of communication efforts'. He further acknowledges that a problem for the empirical operationalisation of arena theory is that since the resources are defined so broadly then 'any social behaviour can be interpreted as a resource mobilisation effort'. Since 'the fundamental axiom is that resource availability determines the degree of influence for shaping policies' this is a serious difficulty.

Nevertheless the arena approach outlined by Renn provides a useful guide to the empirical analysis of risk conflict while accepting his proviso that it needs 'further conceptual and instrumental specification'. The risk arena concept draws on the public arena ideas of Hilgartner and Bosk (1988) yet Renn's treatment makes no reference to the social network tradition on which this drew: 'We borrow from organisational network theory, stressing the influence of and the interrelationships between institutions and social networks in which problem definitions are framed and publicly presented'. Interestingly the accompanying interest in 'the role of drama' echoes the twin concerns of Moreno the pioneer of the network sociogram.

The notion of a 'policy network' has attracted considerable attention in recent years and provides a more general framework within which to locate some of the ideas associated with the arena concept. Thatcher (1998) in a recent review of the field, identifies a central strand as the use of policy network as a generic concept in which to embrace a variety of types of state interest group relations. As well as encompassing a spectrum of the macro political concepts of pluralism and corporatism it embraced more specifically the range of policy subsystems from 'policy community' (Jordan, 1990) to 'issue network' (Heclo, 1978). The new emphasis on policy subsystems flowed from empirical studies which suggested that for many policy areas the general political and legal institutions were not the primary forum in which they were considered but instead interactions occurred within a particular set of state and social actors. The 'policy community' concept tended to emphasise stability and closure in such subsystems while that of 'issue network' suggested fragmentation and diversity.

The evolution of the concept of policy network has been accompanied by the elaboration of criteria on which to base a network typology. These have included integration, membership, distribution of resources, sector boundedness, stability, mobilisation of business interests, autonomy and concentration of the state. The most elaborate includes seven criteria: actors, function, structure, institutionalisation, rules of conduct, power relations, actor strategies (van Waarden, 1992). Many of these studies derive criteria from different theoretical and empirical traditions within political studies and are not explicitly constructed in terms of network concepts.

One group of studies, however, draws on social network analysis to delineate the 'pattern of relationships' in a policy network according to network positions of actors (for example, centrality, strong and weak ties) and/or forms of exchange (for example, information or resources) between actors. Investigative and modelling methods are deployed which enable the 'mapping' of linkages between policy actors and which include, without prejudgement, a diversity of actors (for example, private and public), and of links (for example, formal and informal) (Knoke, 1990; Marin and Mayntz, 1991). The aim of such studies is to analyse the 'structural configuration' of policy networks in order to explain more systematically the conditions within which actors seeks to influence the policy process.

Thatcher (1998), criticises studies within both this and the political studies school for the limitations of their treatment of the relationship between policy network types and either policy processes or outcomes. Borzel too (1997) argues that only a small number of authors have been ambitious enough to seek to attach explanatory value to the different network types i.e., that 'the structure of a network has a major influence on the logic of interaction between the members of the networks, thus affecting both policy process and policy outcome' and observes that 'no hypotheses have been put forward which systematically link the nature of a policy network with the character and outcome of the policy process'. A further criticism directed by Thatcher at both schools is the failure to explain why networks change, alter or arise and what factors, for example, the role of the state, influence the 'distribution of resources among actors, the shape of the networks and the positions occupied by actors within them'. Important issues are the inclusion and exclusion of actors and the rules for their behaviour.

While accepting the greater rigour of social network analysis, Thatcher argues convincingly that it needs to be combined with a wider body of political theory in order to explain the policy process effectively. This is recognised as a difficult undertaking in terms of the identification of coherent and testable propositions and one which requires a more focused and less grand orientation in order to reduce 'the gap between the ambitions and achievements of network approaches'.

Although the policy network approach has attracted criticism for its traditional focus on the stability and continuance of network relationships and its lack of concern with. policy change this has begun to shift significantly in recent years. Sabatier's (1988) 'advocacy coalition framework' has been influential in this. This argues for the importance of studying policy making as a process operating across the various institutions of government which cannot be effectively addressed by the traditional institutional approaches of political science. Stage models of the process are criticised for over simplicity and a top down focus. The ACF framework argues that to understand policy change it is necessary to focus on a policy domain for a period of more than a decade. Actors from all levels of government need to be considered and policies should be conceptualised in a similar way to belief systems. Actors can be aggregated into a number of advocacy coalitions composed of people from various government and private organisations who share a set of normative and causal beliefs and who often act in concert (Jenkins-Smith and Sabatier, 1994).

Visualising Policy Networks

Brandes *et al.* (1999) argue that graphic visualisation of policy networks has considerable value in the furtherance of network analysis and has become more feasible with developments in computer software. They consider that visualisation can facilitate exploration and comparative analysis of network data but this requires a systematic approach to the production of such visualisations based on a definition of explanatory requirements and graphical excellence. The basic goal of the study of policy networks is defined as the 'structural description of the actors and the analysis of relational configurations that are involved in the making of public policies'. Policy network research requires the delineation of the 'set of relevant actors engaged' and the identification of the 'relations among the actors which are of particular significance and consequence for the policy outcome'. The goal is to explain a certain policy by the 'structured interaction within the actor set' with structuring understood as 'an emergent effect which restricts as well as enables'. Two types of structural analysis are pursued: 'connectedness' in terms of the links and flows between actors and 'profiles' of relations in which an actor is involved and their 'equivalence' to other actors. Each of these may be employed at the different levels of actor, group and network and both syntactical and semantic attributes need to be conveyed. It is argued that 'visualisation can enhance the understanding of complex multi-dimensional settings by separating different kinds of information'. Since in policy network analysis there is interest in exploring the different levels of aggregation simultaneously a desirable visualisation technique would 'combine the associated perspectives in an information-rich design that allows us to switch between detail levels within a single image'. The authors make proposals for methods to bring together the three requirements for effective visualisation of policy networks: substance, design and algorithm.

A Research Framework for Exploring Innovation and Risk Networks

Recent work on both innovation and policy networks shows a convergence in approach with a shared interest in capturing a diversity of structural and relational features of networks through visualisation and mapping. Concepts from social network theory such as strong/ weak ties, gatekeeper/boundary spanning roles, network size and density are becoming integrated with organisational and policy concepts such as change management processes and political modes of representation.

The investigation into innovation and risk in the food sector is located in this framework and analyses the actors and their relationships involved in four different generic food innovations associated with visible controversies over risk. These innovations are located at contrasting points in the food system and express different aspects of consumer or environmental risk.

The focal actors in the innovation networks are the organisations which have been directly responsible for the commercial innovations entering the market in the country concerned. These are known as the core innovators. The other network actors are those outside of these core innovator organisations which have made a specific contribution to the development of the innovation. These might include research organisations making knowledge inputs, public bodies offering research grants or regulating the innovation, customers making certain requirements.

The focal actors of the risk network are the policy actors with the power to make decisions on risk. Other members of the network are those organisations or individuals who undertake actions to influence collective social decisions concerning these risks. The actors can be identified through various methods, such as formal representations in the policy process, for example, evidence to a public inquiry, visibility in media coverage of the issue, snowballing methodology after identifying one or two core actors. There is also a wider set of actors, for example, the public, professionals etc. who may have views on the issue and who are the audiences or constituencies to which the core actors may relate in some way.

Through the systematic mapping of these innovation and risk networks it will be possible to compare the configuration of these two types of network and elucidate the relationship between them. Comparisons between subsectors of the food sector, subdomains of risk policy and nations will enable an assessment of the influence of these different settings on network characteristics.

There are two broad propositions in relation to innovation and risk networks in the food sector which merit particular attention. One is that inter-organisational networks have become more important in the innovation process which has changed the traditional influence of large firms. The other is that the policy networks around agriculture and food have changed significantly in form in the past two decades from closed and stable to more diverse and volatile (Smith, 1991). The method adopted will enable answers to be given as to whether such general network characteristics hold up in a such complex sectoral and policy domains.

References

Blauwhof, G. (1994) "Non equilibria dynamics and the sociology of technology," in *Evolutionary Economics and Chaos Theory. New Directions in Technology Studies*, Leydesdorff, L. and P. van der Besselaar (eds.), Pinter, 152–166.

Börzel, T. A. (1997) "What's so special about policy networks? — an exploration of the concept and its usefulness in studying European governance," European Integration online papers, Vol. 1, No. 016.

Brandes, U., P. Kenis, J. Raab, V. Schneider and D. Wagner (1999) "Explorations into the visualisation of policy networks," *Journal of Theoretical Politics*, **11**(1), 75–106.

Callon, M. and B. Latour (1981) "Unscrewing the big Leviathan: How actors macrostructure reality and how sociologists help them to do so," in *Advances in Social Theory and Methodology: Toward an Integration of Micro- and Macro-Sociologies*, K. D. Knorr-Cetina and A. V. Cicourel (eds.), Routledge and Kegan Paul, Boston, Mass., 277–303.

Conway, S. and F. Steward (1998) "Networks and interfaces in Environmental innovation: A comparative study in the UK and Germany," *Journal of High Technology Management Research*, **9**(2), 239–253.

Conway, S. and F. Steward (1998) "Mapping innovation networks," *International Journal of Innovation Management*, **2**(2), 223–254.

Coombs, R. (1995) "Firm strategies and technical choices," in *Managing Technology in Society*, A. Rip, T. J. Misa and J. Schot (eds.), 331–345.

Elzen, B., B. Enserink and W. A. Smit (1996) "Socio-technical networks: How a technology studies approach may help to solve problems related to technical change," *Social Studies of Science*, **26**, 95–141.

Groenewegen, P. and P. Vergragt (1991) "Environmental issues as threats and opportunities for technological innovation," *Technology Analysis and Strategic Management*, **3**(1), 43–55.

Heclo, H. (1978) "Issue networks and the executive establishment," in *The New American Political System*, (ed.) King, Anthony, American Enterprise Institute, Washington DC, 87–124.

Hilgartner, S. and C. L. Bosk (1988) "The rise and fall of social problems: A public arenas model," *American Journal of Sociology*, **94**(1), 53–78.

Irwin, A. and P. Vergragt (1989) "Rethinking the relationship between environmental regulation and industrial innovation: The social negotiation of technical change," *Technology Analysis and Strategic Management*, **1**(1), 57–70.

Jenkins-Smith, H. C. and P. A. Sabatier (1994) " Evaluating the advocacy coalition framework," *Journal of Public Policy*, **14**(2), 175–203.

Jordan, G. (1990) "Subgovernments, policy communities and networks," *Journal of Theoretical Politics*, **2**(3), 319–338.

Kemp, R. (1994) "Technology and the transition to environmental sustainability: The problem of technological regime shifts," *Futures*, **26**(10), 1023–1046.

Kemp, R., J. Schot and R. Hoogma (1998) "Regime shifts to sustainability through the processes of niche formation: The approach of strategic niche management," *Technology Analysis and Strategic Management*, **10**(2), 175–195.

Knoke, D. (1990) *Political Networks. The Structural Perspective*. Cambridge University Press, Cambridge.

Macdonald, S. and C. Williams (1993) "Beyond the boundary: An information perspective on the role of the gatekeeper in the organisation," *Journal of Product Innovation Management*, **10**, 417–427.

Marin, B. and R. Mayntz (eds.) (1991a) *Policy Network: Empirical Evidence and Theoretical Considerations*, Campus Verlag, Frankfurt aM.

Renn, O. (1992) "The social arena concept of risk debates," in *Social Theories of Risk*, S. Krimsky and D. Golding (eds.), Praeger, Westport, 179–196.

Sabatier, P. A. (1988) "An advocacy coalition framework of policy change and the role of policy-oriented learning therein," *Policy Sciences*, **21**, 129–168.

Schot, J., R. Hoogma and B. Elzen, (1994) "Strategies for shifting technological systems: The case of the automobile system," *Futures*, **26**(10), 1060–1076.

Schot, J. W. (1992) "Constructive technology assessment and technology dynamics: The case of clean technologies," *Science, Technology and Human Values*, **17**(1), 36–56.

Smith, M. J. (1991) "From policy community to issue network: Salmonella in eggs and the new politics of food," *Public Administration*, **69**, 235–255.

Steward F. (1995) "Risk analysis and rival technical trajectories: Consumer safety in bread and butter," in *Managing Technology in Society*, Arie Rip, Johan Schot and Tom Misa (eds.), Pinter, 111–136.

Steward, F., S. Conway, S. Yearley, P. Bailey, C. Garcia, A. Hansen, J. Lemarie and J-P Benoit (1998) "Environmental networks and societal management of technological innovation in the food sector," *EC DGXII Environment programme.*

Steward, F. and S. Conway (1998) "Situating discourse in environmental innovation networks: A UK/German Comparative Analysis," *Organization,* **5**(4), 483–506.

Tansey, G. and T. Worsley (1995) *The Food System.* Earthscan.

Thatcher, M. (1998) "The development of policy network analysis," *Journal of Theoretical Politics,* **10**(4), 389–416.

Tidd, J. (1995) "Development of novel products through intraorganisational & interorganisational networks," *Journal of Product Innovation Management,* **12**, 307–322.

Van Waarden, F. (1992) "The historic institutionalisation of typical national patterns in policy networks between state and industry," *European Journal of Political Research,* **21**, 131–162.

Chapter 7

Techno-Economic Networks: Technological Transfer *via* the Teaching Company Scheme

Tim Edwards

Introduction

In this chapter, I discuss the transfer of knowledge and expertise from the knowledge-base to industry *via* the UK government's Teaching Company Scheme (TCS). The main findings suggest that although TCS programmes are organised around core activities and formal management structures, developing innovative solutions remains complex, dynamic and essentially uncertain. This seems to undermine the commonly held belief that the *innovation process* is predictable, linear and manageable (see Saren, 1984; Wolfe, 1994; King and Anderson, 1995). Instead, the evidence presented here supports those studies that describe the innovation process in terms of loosely coupled episodes recursively organised (not sequentially spaced stages) which are subject to conflict, struggle and continual reinterpretation (Robertson, *et al.*, 1997; Clark and Staunton, 1989). In presenting the analysis I draw on two key sources: the sociology of innovation and actor-network theory. In particular, Callon (1986; 1991; 1992) refers to institutional arrangements like TCS as techno-economic networks (TENs) and the process by which they are established and maintained as translation. Conceptualising and analysing TCS from this perspective represents an attempt to make sense of the multiple *social interactions* or *networking* associated with the translation of ideas, through the identification of a problem or opportunity, into reality.

Policy Initiatives for Improving UK Industrial Competitiveness

In the UK, as in other industrialised countries, to a greater or lesser extent, the state has tried to improve the competitiveness of indigenous industries by encouraging links between industry, the knowledge base and government (Freeman and Soete, 1997). At the core of the UK government's economic policy has been a concern with 'building the knowledge driven economy' (OST, 1993; 1998). Although this policy has many different strands including the tight control of inflation and interest rates, the government also makes direct con-tributions to encouraging business competitiveness. To this end, TCS is often cited as the government's premier technology transfer scheme (Quinquennial Review, 1996).

The aim of TCS is to help companies resolve their strategic problems by improving their competitiveness in existing or new markets through the development of products, processes and/or work organisation. TCS facilitates organisational learning through the employment of one or more graduates known as TCS Associates for a minimum of two years (per associate) on company projects jointly supervised by academics and company staff. It is generally accepted that the acquisition of knowledge by individuals does not represent organisational learning instead information and knowledge has to become embedded across the organisation; reflecting a broad mix of individuals, skills and talents (Nonaka and Takeuchi, 1995).

"It is not sufficient for a firm to access useful knowledge. It has also to organise methods for the internal diffusion of new knowledge, to ensure that knowledge which is received from external sources is communicated and utilised effectively throughout the organisation" (Senker and Senker, 1994:81).

This process is particularly difficult if organisations do not have the skilled personnel necessary to implement and guide the process. The strengths of TCS relate to the way an associate and their supervisor help the company identify 'information which can add value to the business' and integrate 'new knowledge into a company's existing store of accumulated knowledge' (Senker and Senker, 1994:82). This is a *learning process* that may require a change of company culture, an important factor in ensuring the continued integration of new knowledge. Associates are a valuable resource in this process, not only because they are subsidised through the TCS grant, but also because they are able to apply 'science and technology as well as modern management and marketing methods properly and systematically'. The associate in turn, benefits from the involvement of the supervisors as the application of knowledge (including tacit) relies on them 'learning through example and experience' (Senker and Senker, 1994:82).

TCS is similar to programmes such as ESPRIT that have been established to improve European technological competitiveness. These programmes are also based on the assumption that competitiveness can be improved if heterogeneous actors including knowledge-based institutions and industry are encouraged to work together to exploit scientific and technological knowledge. Following Laredo and Mustar (1996:145), I propose that such collaborations constitute an 'organisational arrangement crucial to innovation processes: the techno-economic network'.

The Innovation Process

At its most inclusive, the term innovation has been defined to include 'the new markets, the new forms of industrial organisation that capitalist enterprise creates' (Schumpeter, 1943:83) while at its least innovation refers to an object such as a new computer (Swan, 1996). It is now more generally accepted that innovation in organisations represents 'the development and implementation of new ideas by people who over time engage in transactions with others within an institutional order' (Van de Ven *et al.*, 1989:590). Such interpretations reflect a growing interest in the *process* through which 'new ideas, objects, and practices are created and developed or reinvented' (Slappendel, 1996:108). The adoption of process perspectives has facilitated recent attempts to show that organisational innovation involves both social and economic interaction 'encompassing various phases or episodic activities, recursively rather than sequentially related' during which different bodies of knowledge are constructed, communicated and exchanged (Robertson *et al.*, 1997:1–2).

Interest in *process perspectives* indicates disillusionment with the orthodox normative-variance approaches of the 1960s and 1970s (Clark, 1995). The new orthodoxy assumes that a better appreciation of organisational innovation is more likely if attention is given to looking at the nature of, and the different factors affecting, the innovation process. Wolfe (1994) suggests that there have been two generations of process theory research. First, *stage model research* (Pelz, 1983; Ettlie, 1983) which include studies characterising innovation as a series of stages unfolding over time, and second, *process research* (Dean, 1987; Dyer and Page, 1988; Schroeder *et al.*, 1989) which include studies describing the conditions and sequences determining innovation processes. The second version holds particular favour at present because it coincides with current views that innovation is usually a complex iterative process that is not linear and, as such, not best represented by stage models. Saren (1984) provides a review of the types of stage/process models present in the literature. Five models are identified which are defined in terms of those that focus on activities involved in the innovation process; those that focus on activities in relation to specific departments; and those that emphasise the decisions taken in the process:

(i) departmental-stage models;
(ii) activity-stage models;
(iii) decision-stage models;
(iv) conversion process model;
(v) response models.

According to Saren (1984:21), the first three models fail to explain adequately the innovation process because they give insufficient attention to the unexpected. Instead, in an attempt to 'break down the innovation process into its component parts' they inadvertently present 'innovation as an orderly, logical process'. In contrast, conversion process models represent innovation not as a series of steps, stages or phases, rather, the firm is understood to be the user of various types of inputs: raw materials, scientific knowledge and manpower which are converted, in no specified order, into outputs such as products and services. Ironically, this model has been criticised for not aiding attempts to *order* events that constitute the innovation process. The final models which represent innovation in terms of the 'firm's response to some external or internal stimulus' offer only limited insight unless combined with other views (Saren, 1984:23).

Saren's (1984) typology provides an indication of the various approaches to the study of innovation in organisations. Yet, as Saren (1984:24) states, 'despite the number of models and perspectives more work does need to be concentrated on the *nature* of the innovation process in the firm itself' (*my italics*). Although these remarks were made over 15 years ago they remain pertinent today. Developments in this area have come from what might be loosely termed the 'sociology of innovation'. For example, Robertson *et al.*, (1997) propose that our understanding can be improved if attention is given to the variety of roles networks play in the innovation process. Networks comprise process dimensions; defined in terms of the *social activities* undertaken by groups and individuals within and across the firm:

> "Networking involves a search for knowledge and information through the creation and articulation of informal relationships within a context of formal intra/inter organisational relationships" (Robertson *et al.*, 1997:1).

The authors posit that the innovation process consists of various episodic activities (invention, diffusion and implementation) recursively rather than sequentially related through which different bodies of knowledge are constructed, communicated and exchanged (Fig. 1). Each episode constitutes a specific set of activities. The *invention* episode is a personalised process during which knowledge

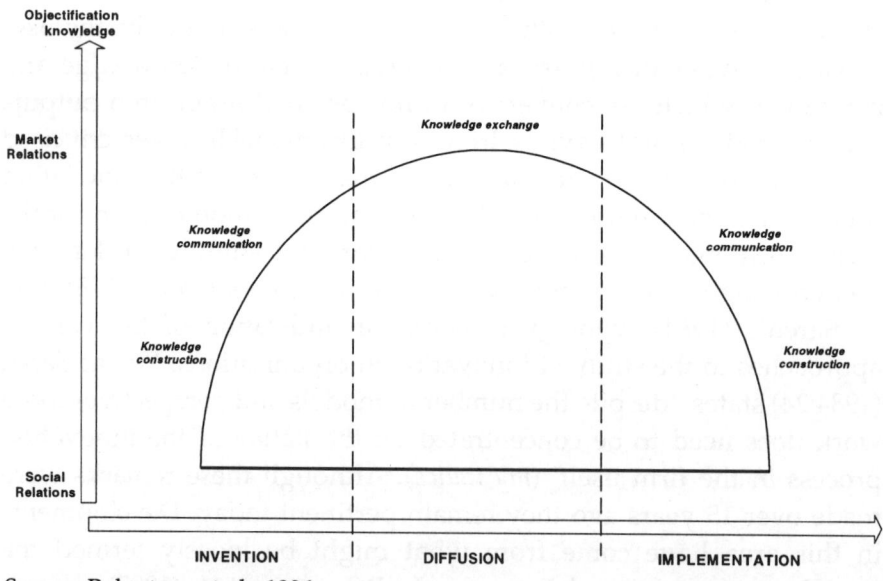

Source: Robertson *et al.*, 1996.

Fig. 1. The innovation process and the role of networks.

is constructed (Bijker *et al.*, 1987). The main objective is to identify potential network participants who posses appropriate skills, expertise and information. Having tapped the 'tacit and contextual knowledge of different individuals and groups' formal and informal teams are assembled (on the basis of uncertain reciprocity and trust) to test and validate the knowledge (Robertson *et al.*, 1997). Parallel to the invention episode is the *diffusion* episode that involves formal and informal exchanges of information among members of the network (Rogers, 1983). Networking activities involve boundary-spanning individuals who translate ideas into locally relevant solutions (Tushman and Scanlan, 1981). Similarly, the *implementation* episode involves the appropriation of knowledge by social groups. In this case, knowledge is constructed to meet the needs of the particular context. The appropriation of knowledge consists of periods of construction and communication between individuals and social groups who are engaged in 'fitting' the knowledge within the organisation (Clark, 1987).

Linking networking activities with the innovation process has generally been over-looked with exceptions that include Rogers (1983), Alter and Hague (1993), Conway (1994) and Robertson *et al.* (1997). Where networking has been applied it represents an important development in explaining the innovation process. Key to this is the identification of networking activities that organise social practices and translate ideas into reality (Knights *et al.*, 1993). Scholarly interest in such processes is apparent in a number of areas including: the social construction of technology (Bijker *et al.*, 1987); the social shaping of technology (Edge, 1988; Mackenzie, 1991; Mackenzie and Wajcman, 1985; Williams and Edge, 1991); and actor-network theory or the sociology of translation (Bijker and Law, 1992; Callon, 1986; 1991; 1992; Latour, 1987; 1988; Law, 1988; 1991). In this study, I draw explicitly on actor-network theory in an attempt to open the *black box* of innovation to sociological analysis.

The Network Perspective

Network perspectives have proven useful for expressing relational data amongst objects such as people, groups and organisations joined by a variety of relationships (Tichy *et al.*, 1979). Fombrun (1982:280–281) defines networks as 'a powerful means of describing and analysing sets of units by focusing explicitly on their inter-relationships' which are 'embedded in a context that both constrains and liberates'. Such a view is based on the sociological tradition which has attempted to investigate and identify the causes and consequences of the structure and patterning of relationships in social systems (Marsden and Lin, 1982; Scott, 1991; Tichy *et al.*, 1979). The different research fields adopting this approach include, for example, the public sector (Kickert *et al.*, 1997), the diffusion of innovation (Rogers and Shoemaker, 1971; Robertson *et al.*, 1996), the innovation process (Robertson *et al.*, 1997), and knowledge work (Knights *et al.*, 1993).

As with other sociological traditions the network perspective is characterised by different assumptions about the social world that have resulted in a pluralism of theories, raising the question of mutual incompatibility or complementarity (Van Poucke, 1980). Differences in network approaches reflect the various ways authors have attempted to clarify the term. A distinction can be made between those studies drawing on a metaphorical orientation and those that use a mathematical orientation (Conway, 1994). The network as metaphor represents a way of *picturing* the structure of social reality: an organism in relation to its environment. In contrast, those adopting a mathe-matical orientation set out to classify 'the sorts of bonds between individuals and... the patterns formed by them, to discover what causal connections there may be between the various patterns and the behaviour of the individuals belonging to the networks' (Van Poucke, 1980:181). Network studies also reflect the relational and structural research traditions (Conway, 1994). Relational network analysis concentrates 'on the pathways in networks and entails identifying the cliques of individuals among the members of a network' (Conway, 1994:73). This perspective 'attempts to gather insights into how actors manipulate networks to reach certain aims' (Van Poucke, 1980:182). In contrast, the structural perspective traces the role of network properties on the behaviour of individuals and organisations (Bott, 1971; Laumann, 1973).

Adopting mathematical/structural or metaphorical/relational perspectives has tended to perpetuate the philosophical division between determinist and voluntarist orientations in social analysis. In contrast, the actor-network approach used here attempts to resist determinist tendencies that assume networks have inherent characteristics. This is achieved by denying the essentialist or *a priori* analytical distinction between the social and technical (Grint and Woolgar, 1997). Instead, there is focus on alignments between the social and technical elements as they are *enrolled* into an innovation network. Establishing techno-economic networks (TENs) involves the utilisation of a set of intermediaries that give material content to the links uniting actors (Callon, 1986). Hence, the analysis focuses on the way human agents struggle with one another to 'first determine their existence and then (if that is secured) define their characteristics... the outcomes of these struggles depend upon the particular combination of elements in play' (Law, 1986:15–16). The key assumption is that power is the outcome of a set of strategies rather than being possessed by individual agents. Consequently, strategies reflect the interplay of new and pre-existing institutional forms, arrangements and interpretations: an agent's actions are never totally voluntary.

The Innovation Process and Actor-Network Theory

Techno-economic networks are defined as 'a co-ordinated set of heterogeneous actors which can include public laboratories, research centres, companies, financial institutions, government, and users' (Callon, 1992:73). Various actors participate collectively in the conception, development, production, distribution and diffusion of procedures for producing goods and services. TENs are organised around three poles: the scientific pole which produces empirical knowledge and consists of universities and other independent research centres; the technical pole which comprises technical laboratories, pilot plants, development engineers and scientists, which utilise the empirical knowledge for prototypes, models, tests and trials; and the market pole which contains users who generate specific demands or needs, and attempt to fulfil them (Callon, 1991). In addition to these three *supporting pillars* there are two mediating poles. The commercialisation pole 'consists of production and distribution activities that mobilise technology to create/satisfy needs' and the transfer pole which 'specialises in connecting science and technology' (Laredo and Mustar, 1996:159).

> "The processes of production and exchange that we can observe taking place in TENs involve a whole series of intermediation between these poles.... The different poles have memberships, goals and procedures which may apparently be mutually exclusive...however, arrangements and links are made between the members of different poles, so that the outputs of various activities are exchanged with the members of other poles" (Callon, 1992:74).

The establishment and evolvement of TENs emerges from interaction between the social and the scientific/technological realms which involves a number of diverse actors located around these poles. Interaction is, in turn, dependent on intermediaries that pass between actors *'which defines the relationship between them'*. Consequently, the relationship between actors and the intermediaries mobilised during interaction is irrevocably tied, such that, *'actors define one another in interaction — in the intermediaries they put into circulation'* (Callon, 1991:135, italics in original). Callon (1991; 1992) refers to four categories of intermediary when defining the relationship between actors:

texts or literary inscriptions include anything that is written, such as advertisements, reports or software. *Technical artefacts* are non-human entities that facilitate functions and tasks such as software, prototypes, machines etc. The *skills* of human beings and the knowledge they generate and reproduce. Finally, *money* in a multiplicity of forms including subsidies, grants and loans. TENs usually consist of combinations or hybrids of intermediaries: human and non-human, individual and collective entities.

In considering the make-up of TENs it is important to distinguish between 'actor' and 'intermediary'. This distinction gives notice that all interactions reflect some form of authorship: '*An actor is an intermediary that puts other intermediaries into circulation*' (Callon, 1991:141, italics in original). Actors are distinguished from intermediaries because they, as authors, draw upon methods for imputing intermediaries to a particular relationship:

> "...an actor is an entity that takes the last generation of intermediaries and transforms (combines, mixes, concatenates, degrades, computes, anticipates) these to create the next generation. Scientists transform texts, experimental apparatus and grants into new texts. Companies combine machines and embodied skills into goods and consumers. In general then, actors are those who conceive, elaborate, circulate, emit, or pension off intermediaries" (Callon, 1991:141).

Although the process through which actors attempt to define identities as well as relationships between themselves and other intermediaries is uncertain and controversial it does represent the evolvement of a set of equivalences and alliances around a particular scenario which carries the *signature of its author*. In this way, an actor is also a network (hence actor-network) that is actively seeking other entities to embark upon a set of clearly defined relationships (Callon, 1991).

The emergence of a TEN around a clear set of equivalences and definitions is a complex endeavour that Callon explores through the *process of translation.* Translation involves a number of overlapping dimensions: *problematisation, interessement, enrolment* and *mobilisation.* *Problematisation* involves one or more actors and/or institutions working together to define and explore a problem. *Interessement* involves actors who believe they have a relevant solution to a problem persuading others to agree with their definition and collaborate to find the favoured solution. *Enrolment* involves the establishment of various co-ordinating mechanisms and procedures to build and maintain the running of the network. While, *mobilisation* involves employing a range of rules and methods to sustain the network (Callon, 1986; 1991; 1992). In this way, 'a concern with translation focuses on the process of mutual definition and inscription' (Callon, 1991:143), through which the interplay of actors and intermediaries are ordered.

Methodology

Network analysis involves a number of decisions concerning the scope and content of analysis. According to Mitchell (1969), the researcher should decide what aspect of the *total network* is the subject of investigation. This is significant because 'the general ever-ramifying, ever-reticulating set of linkages that stretches within and beyond the confines of any community or organisation' is usually too vast to study in its totality (Mitchell, 1969:12). Hence, it is necessary to identify a *partial network* and this is dependent on a process of abstraction (Scott, 1991). Abstraction begins by deciding the definitional focus of the study (Laumann *et al.*, 1983) which relates to three components: (i) actors, (ii) relations and (iii) activities. One or more of these sets can be used to establish the partial network. The first definitional focus is based on finding a common element that links actors in a network. The second definitional focus is dependent on the participation of actors in some specified social exchange. The third definitional focus relates to an event or activity that serves to select individual actors and the social relationships amongst them. This method can be used to construct action-sets which are special kinds of networks established to realise some preferred objective (Conway, 1997). Abstraction also depends on deciding how the partial network is anchored or centred. A network may be anchored around a particular actor, referred to as *ego-centred* networks, or a population of actors, known as *socio-centred* networks (Conway, 1997; Scott, 1991).

For the purposes of defining the scope of the current analysis an action-set approach based on the TCS programme's tasks seems most appropriate. In conjunction, a socio-centred nodal anchoring approach is used as the TCS programme constitutes a special set of social relations amongst a number of key actors: the Teaching Company Directorate (TCD) consultant, the TCS centre manager, the associate and academic and industrial supervisors. The empirical data are derived from a longitudinal case study covering 20 months (May 1997–Dec 1998). This detailed qualitative investigation involved a TCS programme in a small firm employing 25 people manufacturing and distributing a range of cleaning products for computers and other office equipment. Beta, as the company is called for reasons of confidentiality, collaborated with a University Business School to 'develop and implement a sophisticated modern purchasing management system within a TQM philosophy appropriate to the rapidly expanding and increasingly international market place' (TCS, 1997:2). The data are taken from documents such as reports (×2), letters and memos (×9), minutes of project management meetings (×1), minutes of local management committee meetings (×4), in-depth interviews (×12) and observation including attendance at meetings with the main participants (×6). Additional data have also been obtained from the TCS centre manager, Birmingham and the TCD consultant monitoring the programme.

In this chapter, I am concerned to critically analyse the process of turning ideas into reality through the creation and running of a TCS programme. Transforming discourse into action is a difficult task relying as it does on the coordination of disparate entities human and non-human. Critiquing this highly political process using actor-network theory has two main advantages: first, the notion of translation helps to clarify the methods and discourse associated with the construction of technological alliances and, secondly, it provides insight into the nature of the fault lines that pervade these temporary collaborations. This offers an alternative to normative linear models of innovation that tend to be uncritical simplifications of complex processes.

The Beta Teaching Company Scheme:
A Techno-Economic Network

TCS programmes link the scientific and market poles through the development of the technical pole (see Fig. 2). Linking these poles relies on reciprocal interrelations between the associate and their supervisors organised around the mediating functions of the transfer and commercialisation poles. It is during this process that 'the identity of actors, the possibility of interaction and the margins of manoeuvre are negotiated and delimited' (Callon, 1986:203). To begin, the author

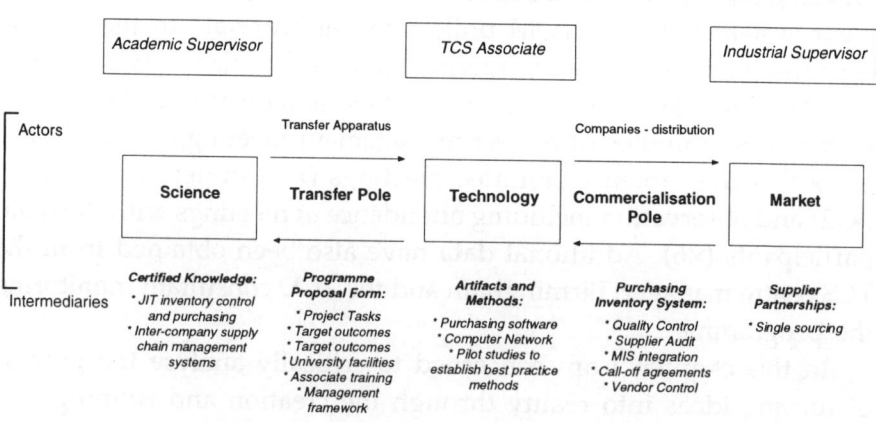

Adapted from Laredo and Mustar, 1996.

Fig. 2. The Techno-Economic Network.

outlines the way in which the alliance of intermediaries and actors were first defined and then consolidated. Consensus is never guaranteed as actors continually re-assess the validity of such alliances. The translation of ideas may be contested and open to struggle, on such occasions alliances may break and new solutions preferred.

At Beta, as with any TCS programme the organisational arrangements centred on the activities of the associate. In this case, the associate was set a detailed work schedule to *develop and implement a new and sophisticated purchasing management system*. The decision to do so originated with the Commercial Director (industrial supervisor) who saw a need to improve Beta's functional integration, the monitoring of suppliers and response to customers (the market pole). Success required the associate accessing (developing the technical pole) the academic supervisor's practical knowledge and experience. Which, in this case, was in just-in-time inventory control, purchasing and inter-company supply chain management (the science pole). These activities and interdependencies were organised around a set of *tasks* in the associate's work schedule. Establishing and activating these tasks depended on dialogue between the actors so that their alliances and responsibilities remained suitably tuned to developing and implementing the new purchasing system.

The Four Moments of Translation

Problematisation

Problematisation originated with the experiences of senior management at Beta in the six months prior to talks about TCS funding (late 1995, early 1996). At this point, Beta merged with another company and the managing director was appointed commercial director. After a short period in post, the new commercial director decided that the company lacked the management systems to operate effectively (compounded by Beta doubling in size as a result of the merger). Beta's managing director had, until the merger, been unwilling to devolve responsibility or to adopt professional administrative procedures. According to the commercial director: 'Virtually everything was done on the back of a fag packet'. For example, the purchasing manager did not keep accurate records nor did he monitor suppliers. Until the start of the TCS programme the purchase of raw materials had taken place without due consideration of the 'options available and without balancing price against stock cost' (TCS, 1997:3). Contact with the TCS centre manager coincided with the commercial director's attempts to implement changes at Beta. These included the implementation of improved methods to keep accurate account of orders and dispatches, the appointment of a new marketing director to enhance the marketing function and building better linkages between marketing, sales, manufacturing and purchasing.

Problematisation is concerned with *recruiting* and *defining* a set of actors, which in this case were involved in the Beta TCS. The *inter-definition of the actors*, a term coined by Callon (1986) formally began with the commercial director making an enquiry to Birmingham TCS Centre. The TCS centre manager provided an *initial presentation* outlining participants' supervisory and financial commitments as well as the business benefits that could be expected during a programme (for example, increased profits). This presentation also provided the TCS centre manager with an opportunity to judge whether TCS was the most suitable mechanism to assist Beta. A short while after the initial presentation, the commercial director re-contacted the TCS centre manager to pursue the application for funding. At this point, the TCS centre manager provided an *awareness session* that had three objectives. First, to formally outline the eligibility criteria, timescales and procedures for obtaining funding, secondly, to establish whether the company fitted the eligibility criteria and, thirdly, to develop a one-page brief to outline the proposed TCS programme including actors, objectives, tasks and deliverables. During this session, the commercial director outlined his vision of the programme's objectives and these were incorporated into the one-page brief. He also provided company details including its most recent annual and management accounts. All documents were passed-on to the TCD consultant who made an initial judgement about the eligibility of the proposal and commercial viability of the company. In addition, the centre manager and TCD consultant began to consider which academics would meet the requirements of the proposed programme. They then contacted a Business School academic who after considering the outline proposal agreed to participate in drawing-up a detailed proposal.

The one-page brief established the basic objectives and helped create a network, which linked Beta's commercial director, representatives of TCS and an appropriate academic. At this stage, although the proposal lacked specific detail about the tasks of actors and intermediaries it did indicate an obligatory passage point (Callon, 1986) (the development and implementation of a new and sophisticated purchasing system) through which the various actors would define their responsibilities. Although problematisation starts with these initial negotiations it does not have a clear cut-off point and it continues parallel to the interessement stage. As the academic became more involved with the programme he embarked on negotiations to make his role more significant. In doing so, these actors utilising several interessement mechanisms began locking themselves into a range of alliances.

Interessement

Interessement involved *instituting* the *associations* between the commercial director and academics as well as actors and intermediaries (for example, the associate) yet to be enrolled in the TEN (Callon, 1986). At the point of the awareness session, the envisaged interrelationships between the commercial director, academic and other potential intermediaries (including the associate) were still to be formally negotiated and agreed. Hence, interessement was marked in the first instance by the TCD consultant and TCS centre manager ensuring that the academic and industrial supervisors were aware of their administrative duties in submitting a formal request for funding. Secondly, (reflecting a continuation of problematisation) making sure each party agreed on a solution to the problem of matching business needs academic expertise and technical targets. To this end, both objectives (and problematisation) depended on several interessement mechanisms including texts, meetings, visits and the TCD selection criteria.

Negotiations concerning these roles and associations began at the *partners' initial meeting* and were formally measured against TCD selection criteria. Progress in forming a solution to the business problem depended in the first instance on meeting these criteria thereby defining the actors according to a set of key assumptions relating to their eligibility. Hence, the commercial director had to provide evidence that Beta was (a) viable and anxious to improve its future business performance and (b) had identified those areas it lacked expertise and required assistance. Furthermore, these were then represented as tangible benefits including increased profits as well as qualitative benefits such as improved internal capabilities and staff development. The commercial director also had to show how these benefits were to be consolidated when the programme ended. The academic partner was required to justify his involvement with reference to the benefits the programme would bring to them and their institution in terms of both teaching and research. In addition, the academic was expected to verify the skills, techniques to be drawn upon during the programme and to provide a clear description of the technology to be transferred and embedded into the company.

These associations and responsibilities were formalised with the drafting of the grant application and proposal form (see enrolment). They also represented a continuation of problematisation because each actor became locked into a long line of associations defined according to project objectives, target outcomes and task sequences. From an administrative perspective the TCD consultant and TCS centre manager were responsible for imposing and stabilising the identities and duties of the industrial supervisor and academic supervisor for the purposes of applying for funding. In addition, they engaged with the supervisors in discussions concerning resolution of the business problem. This included identifying the role and activities of the associate. Thus, interessement included TCS selection processes and the organisation of responsibilities around problematisation (the development of a new purchasing system). Such activities led seamlessly on to the enrolment of actors achieved *via* a number of co-ordinating mechanisms and procedures that established alliances according to a set of tasks.

Enrolment

Enrolment involved the commercial director, academic and associate *negotiating* (specify, convince, or coerce others around the problematisation) their *identities* and *roles*. The issue here was to transform the one-page brief into a series of more specific statements. Enrolment depended, at this early stage, on the writing of the grant application and proposal form thereby anchoring interessements and problematisation activties. In writing the grant application and proposal form the academic (in this case) locked himself, the industrial supervisor and the associate into a set of mutual alliances including intermediaries (for example, technical artefacts) that had yet to be formally introduced. As the arrangements were established the academic (with assistance from the TCD consultant and TCS centre manager) utilised a number of strategies to enable them to define responsibilities; in this case according to TCD guidelines and specific work tasks.

To begin, the commercial director and academic were enrolled by virtue of their formal agreement to participate in the programme according to TCD guidelines, which was legally binding. In addition, enrolment involved the allocation of responsibilities between the supervisors and associate. At this stage, these transactions were based on general notions of *mutuality*: the commercial manager resolved his business problem, the academic had the opportunity to apply his knowledge in a business environment and the associate received on-the-job training as well as being equipped for future employment at the host company.

> "To describe enrolment is thus to describe the group of multilateral negotiations, trials of strength and tricks that accompany the interessements and enable them to succeed" (Callon, 1986:211).

In all TCS Programmes, clear guidelines were provided by TCD about expected responsibilities on the project. The academic was expected to provide expertise, skills and aid access to university facilities. The company supervisor (commercial director) was tasked with monitoring the associate's day-to-day work ensuring that they understood the requirements of the company. At a more detailed level, the various actors were also responsible for the accomplishment of a number of interrelated work tasks that had been written into the grant application and proposal form. Although the commercial director and academic discussed the original definition of the business problem (outlined in the one-page brief) it was the academic who translated the *problem* into a work schedule including the introduction of new processes and techniques (some of which would be computerised). This was achieved without the commercial director's direct involvement because he perceived the academic as the expert and left such detail to his judgement. Hence, the commercial director ceded control of creating the programme's architecture to the academic. This was not unrealistic because the academic had the skills and expertise (the science pole) to resolve Beta's business problems. Consequently, the academic took final and as it turned out unmediated responsibility for enrolling the various intermediaries around the original problematisation and interessement.

The establishment of the Beta TCS programme illustrates the different ways actors can be enrolled including: transaction, statutory enforcement, mutuality, allocation and consent by delegation. These derived from multilateral negotiations that transformed the one-page brief into clear statements of intent and responsibility (the TCS grant application and proposal form). In time, however, these identities, alliances and responsibilities which took several months to establish were renegotiated or neglected as the actor-network became subject to changing circumstances and mutuality was replaced with mistrust. Gradually the commercial director appropriated authorship of the network and excluded the academic supervisor. Such actions reflect additional enrolment strategies including 'marginalisation'. Formally, the academic did not challenge the dissipation of alliances, which seemed to reflect a reluctance or inability to resolve the contentions that had influenced these strategies (see below).

Mobilisation

Mobilisation occurs when a number of *rules* and *methods* are employed to *sustain* a network. According to Callon (1986:216): 'To mobilise...is to render entities mobile which were not so beforehand'. At Beta, the commercial director, academic and the associate were the first to be mobilised around a set of equivalences but as the programme progressed, other intermediaries were introduced. For example, as part of the development of the new purchasing system the associate was responsible for auditing and monitoring suppliers. To do so, it was necessary to deploy a prototype monitoring system consisting of a number of proformas. This allowed the associate to begin building a historical record that could, in time, be used to assess and select preferred suppliers. Hence, development of the pro-formas represented mobilisation by virtue of the methods introduced by the academic: (i) the development of a professional infrastructure in the purchasing function; (ii) provision of a high degree of control and sophisticated management information; (iii) rationalisation of the company's supplier base, purchasing routines and controls; (iv) reduction in purchasing costs, lead times and stockholding.

The mobilisation of activities around associated tasks was designed to ensure that the interests of all the participants were met. All actors and intermediaries (for example, the proforma) were assembled at a particular time and place in accordance to a series of displacements (Law, 1985) that framed them and their responsibilities. These activities included formal and informal exchanges of information between members of the innovation network (diffusion) that were mobilised (including the proformas) to fulfil the expectations of Beta's senior management (implementation). These responsibilities were expected to be sedimented by a system of formal management procedures. This system was implemented to ensure that the arrangements specified in the grant application and proposal form was fulfilled and appropriately signed-off. At Beta, formal arrangements included weekly contact between academic and associate, daily contact between industrial supervisor and associate, joint meetings between all three (supposedly monthly) and four-monthly Local Management Committee (LMC) meetings which included the TCD consultant discussing the programme progress, future plans, and budget.

The grant application and proposal form provided a 'project recipe' of rules, management systems, administrative arrangements and formal decision making mechanisms to organise the programme. In reality, translation was far more problematic once the respective actors embarked upon enacting the recipe. As it turned out, the alliances and responsibilities defined at the early stages of the TCS programme (pre-funding) shifted following the appointment of the associate. This can be illustrated by showing how the mobilisation mechanisms (LMC meetings) were used to cover-up the schisms that emerged in subsequent months. Having started the programme it was necessary for the supervisors and associate to ensure that the TCD consultant was 'shown' that programme targets and milestones were being met. This was to be achieved through performance indicators that illustrated savings profits and benefits accrued during the programme. Such benefits were presented at LMC Meetings with reports from each participant on progress (matching tasks with benefits) including any decisions about future changes to the work. On the surface, all appeared harmonious during such meetings as the TCD consultant was given tangible evidence of the programme's 'success'. In this instance, the figures provided little clue to the increasingly dysfunctional state of network relations. If these social relations had become the subjects of discussion then the validity of the alliances would have been open to discussion.

Fragmenting the Network

The analysis has so far described the formation of what were in effect a set of *latent* alliances between the management team at Beta and academics in a Business School. This was formally agreed with the writing and submission of grant application and proposal forms. In being awarded a grant these actors and, subsequently, the associate agreed to collaborate on resolving Beta's business problem. In doing so, these actors had specified an obligatory passage point although negotiations continued well after approval of the grant. Receiving approval meant enrolling an associate into the network as a means of delivering tangible business solutions.

Commencement of the programme at Beta was quickly followed by controversy. At the time of submitting the grant application the academic was told that the associate would work with the purchasing manager to develop a new system. This was an important consideration because the purchasing manager was responsible for the day-to-day running of the function and this would leave the associate free to undertake the evaluative and developmental work. However, before completion of the grant application senior management dismissed the purchasing manager. More significantly, the decision not to replace him only came to light after the associate's appointment. As a result, the associate became responsible for both day-to-day purchasing as well as the programme itself.

By this action, Beta's management team revised the dynamics of the network. It was now impossible for the associate to devote all her time to the programme's tasks. The associate, who in her own words was a 'purchasing virgin' was placed under immense pressure to fulfil the role of purchasing manager and deliver on the programme. Reconciling these demands was beyond the associate as demonstrated during the first LMC when the industrial supervisor (commercial manager) expressed his concern about her 'ability to do the job'. Furthermore, he made it clear that the associate's first responsibility was the purchasing role and if this was not carried out satisfactorily he would be forced to dismiss her. This not only placed the associate in an impossible position it also presented the academic supervisor with a dilemma. He now had to decide how far he should become involved in the associate's day-to-day purchasing activities. Given the associate's lack of experience she, and the commercial manager, now saw the academic's role as that of purchasing mentor. However, with the academic's other Business School responsibilities he was reluctant to become closely involved with the running of the company. By remaining 'distant' from the everyday activities both the associate and Beta's management team soon become disillusioned with the academic's seeming lack of involvement.

This proved a crucial juncture in the mobilisation of the network because it created conflict between the academic supervisor and other participants. It was perhaps less important that these expectations were unrealistic, rather what was significant were the repercussions from this episode. The academic was increasingly excluded from company discussions about the purchasing system and the role of the associate. For example, only two out of the planned 20 monthly management meetings ever took place during the programme. At the second (which turned out to be the last) meeting the industrial supervisor asked the academic whether 'we are keeping you awake' during a discussion about problems they were having with suppliers. The significance of this comment should not be overly interpreted but it does indicate a shift in the perceived and actual roles of the network participants. This is supported by the associate's increased reluctance to request help and advice from the academic. There appeared to be a number of reasons but loss of credibility (reflected in the last monthly meeting) was an important contributory factor.

The effect of these developments was to gradually shift the focus of the solution away from the academic's problematisation towards the Beta management team. Hence, efforts to improve the purchasing system came to reflect co-authorship between associate and the industrial supervisor. The most important impact of this co-authorship was the 'downgrading' of the programme's tasks and objectives. The notion of a 'sophisticated' system was effectively re-written as an 'adequate' system built to the specifications of the industrial supervisor and not the academic. However, it would be wrong to suggest that the programme did, as a result of these re-alignments, fail to deliver a new purchasing system or that the academic supervisor did not provide valuable input. To date the associate has succeeded in installing a whole range of procedures and routines (under the tutelage of both supervisors) which were not present before her appointment. Nevertheless, the new purchasing system does not meet *all* the specifications made by the academic supervisor. Perhaps, this 'downgrading' reflects, what Callon (1986:218–219) terms *treason*.

"If consensus is achieved, the margins of manoeuvre of each entity will then be tightly delimited... But this consensus and the alliances which it implies can be contested at any moment. Translation becomes treason."

Changes to the programme were tabled at each of the LMC meeting so that the TCD consultant was aware of developments in the associate's work plans. However, these changes were usually presented as a *fait accompli*. In doing so, the mechanisms intended to facilitate healthy discussion (for example, LMCs) were actually used to stifle dissenting voices. Such meetings offered little or no insight into shifts in problematisation because the political activities occurred outside the formal arena. This is not to deny the academic supervisor nor any one else could have tabled their concerns during an LMC. In fact, this happened at the first LMC when it became apparent that the associate was having trouble carrying out the purchasing manager's function. However, it appeared that any attempt to 'force the hand' of the industrial supervisor would have worked against the academic in the long term. Throughout the programme it was clear that one of the academic's main concerns was to ensure eligibility for other projects via TCS. It seems probable that failure at Beta, the Business School's first TCS, may have been detrimental to any future aspirations in this direction.

Summary

Establishing and running a TCS programme reflects social relations between three main poles, scientific, technical and market, as well as two mediating poles: commercial and transfer. Representatives of the three poles, the academic and company supervisors and the associate were mobilised to achieve a set of tasks to develop and implement new knowledge and technology according to the objectives and target outcomes of the proposed programme. In doing so, the participants embark on a set of alliances which are established, maintained and perhaps re-negotiated through the various moments of translation. The notion of translation is based on a continual process that has no end point. As long as a TCS programme is running actors will be involved in redefining the problem and negotiating alliances as has been demonstrated in the case at Beta. In doing so, the participants depend on a number of intermediaries including texts such as the grant application/proposal form on which the successful application depends and the monitoring proformas. Skills, which initially include those of the TCS centre manager, TCD consultant and later those of the supervisors and the associate. In Beta these skills, shared between the participants so that knowledge was transferred and embedded, were organised around the implementation of a purchasing system. Finally, money in the form of the TCS grant and company contributions were used to finance the project which included new purchasing software (technical artefacts) as well as payment of the associate and the academic.

The moments of problematisation, interessement, enrolment and mobilisation reflect those activities that incorporate and organise the construction, communication and exchange of knowledge (for example, those activities associated with the introduction of the monitoring proformas). Although the nature of those alignments may be constantly revised and openly contested as at Beta, it is apparent that each moment of the process of translation encapsulates and orders the various activities associated with the innovation process (Fig. 3). Translation provides a means to appreciate reinvention and complexity in the innovation process, where innovation is perhaps more likely to reflect the interplay of conflicting interpretations and power

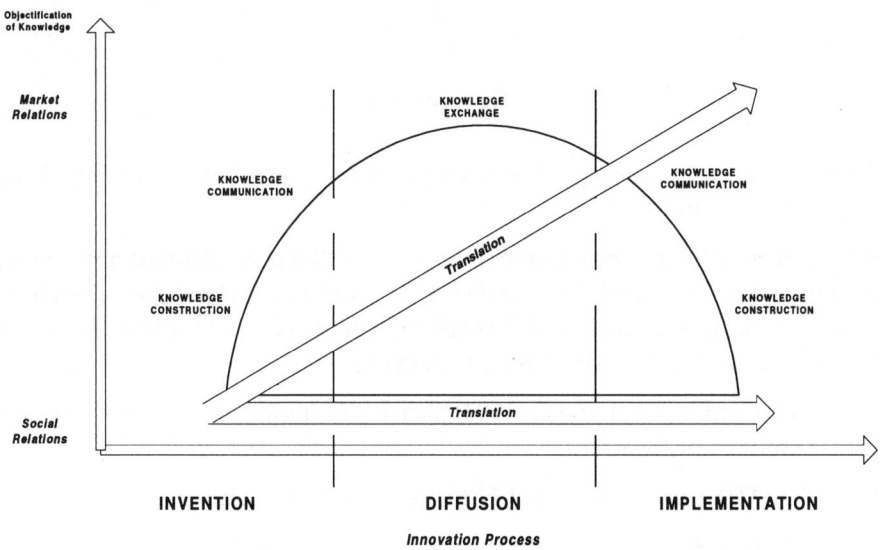

Fig. 3. The innovation process and the sociology of 'translation'.

struggles than rational choice. As Callon (1986:211) states: 'No matter how...convincing the argument...the device of interessement...[like any other dimension]...does not necessarily lead to alliances' or if it does these can change.

Utilising actor-network theory in this analysis of the Beta TCS programme offers an alternative perspective to process models of the innovation process reviewed above (see Saren, 1984). Instead of ordering the activities around a clear sequence of activities (thus inadvertently over-simplifying the process) actor-network theory focuses attention on innovation as a 'translation' by which discourse is turned into reality. Hence, the objective is not to present innovation as a linear process (although for the sake of clarity the analysis considered each element of translation separately) but to explore the strategic conduct of actors (whether human or non-human) in establishing and maintaining collaborative alliances. Identifying the mechanisms through which knowledge is negotiated provides a critical overview and interpretation of the nature of the uncertainties usually associated with innovatory activity.

References

Alter, C. and Hague, J. (1993) *Organisations Working Together*. Sage Publications, London.

Bijker, W. E. (1987) "The social construction of Bakelite: Toward a theory of invention," in *The Social Construction of Technological Systems: New Directions in the Sociology and History of Technology*, Bijker, W. E., Hughes, T. P. and Pinch, T. J. (eds.), The MIT Press, Cambridge.

Bijker, W. E. and Law, J. (eds.) (1992) *Shaping Technology/Building Society*. Mit Press, Cambridge, Mass.

Bott, E. (1971) *Family and Social Network*. Tavistock, London.

Burns, T. and Stalker, G. M. (1961) *The Management of Innovation*. Tavistock, London.

Callon, M. (1986) "Some elements of a sociology of translation: Domestication of the scallops and the fisherman of St Brieuc Bay," in *Power, Action and Belief: A New Sociology of Knowledge?* Law, J. (ed.), Routledge and Keegan, London.

Callon, M. (1991) "Techno-economic networks and irreversibility," in *A Sociology of Monsters: Essays on Power, Technology and Domination*, Law J. (ed.), Routledge and Keegan, London.

Callon, M. (1992) "The dynamics of techno-economic networks," in *Technological Change and Company Strategies: Economic and Sociological Perspectives*, Coombs, R., Richards, A., Saviotti, P. P. and Walsh, V. (eds.), Harcourt Brace Janovich, London.

Clark, P. A. (1987) *Anglo-American Innovation*. Walter de Gruyter Berlin, New York.

Clark, P. A. (1995) "Technical systems and organisational innovations: Duality and the supplier-user junction," *Workshop, Shaping of Technology, IAMOT*, Aston University, July.

Clark, P. A. and Staunton, N. (1989) *Innovation in Technology and Organisation*. Routledge, London.

Conway, S. (1994) "Informal boundary-spanning links and networks in successful Technological Innovation," unpublished Ph.D., Aston Business School.

Conway, S. (1995) "Informal boundary-spanning communication in the innovation process," *Technology Analysis and Strategic Management*, 7(3), 327–342.

Conway, S. (1997) "Focal innovation action-sets: A methodological approach for mapping innovation networks," Research Paper Series, No. RP9702, Aston Business School.

Dean, J. W. Jr. (1987) "Building the future: The justification process for new technology," in *New Technology as Organisational Innovation*, Pennings, J. M. and Buitendam, A. (eds.), Ballinger, Cambridge Mass., 35–58.

Dyer, W. G. Jr., and Page, R. A. Jr. (1988) "The politics of innovation," *Knowledge in Society: An International Journal of Knowledge Transfer*, 1, 23–41.

Edge, D. (1988) "The social shaping of technology," University of Edinburgh PICT Working Paper, No. 1, Edinburgh.

Ettlie, J. E. (1983) "Organisational policy and innovation among suppliers to the food processing sector," *Academy of Management Journal*, **26**, 113–29.

Fombrun, C. (1982) "Strategies for network research in organisations." *Academy of Management Review*, **7**(2), 280–291.

Grint, K. and Woolgar, S. (1997) *The Machine at Work: Technology, Work and Organisation*. Polity Press, Cambridge.

Kickert, W. J. M., Klijn, E-R. and Koppenjan, F. M. (1997) *Managing Complex Networks: Strategies for the Public Sector*. Sage Publications, London.

King, N. and Anderson, N. (1995) *Innovation and Change in Organizations*. Routledge, London.

Knights, D., Murray, F. and Willmott, H. (1993) "Networking as knowledge work: A study of strategic interorganizational development in the financial services industry," *Journal of Management Studies*, **30**(6), 975–996.

Laredo, P. and Mustar, P. (1996) "The technoeconomic network: A socioeconomic approach to state intervention in innovation," in *Technological Collaboration: The Dynamics of Cooperation in Industrial Innovation*, Coombs, R., Richards, A., Saviotti, P. P. and Walsh, V. (eds.), Harcourt Brace Janovich, London.

Latour, B. (1987) *Science in Action*. Open University Press, Milton Keynes.

Latour, B. (1988) "The prince for machines as well as for machinations," in *Technology and Social Process*, Elliot, B. (ed.), Edinburgh University Press, Edinburgh.

Laumann, E. O. (1973) *Bonds of Pluralism: The Form and Substance of Urban Society*. Wiley, New York.

Laumann, E. O., Marsden, P. and Prensky, D. (1983) "The boundary specification problem in network analysis." in *Applied Network Analysis: A Methodological Introduction*, Burt, R. and Minor, M. (eds.), Sage, Beverly Hills, 18–34.

Law, J. (1985) "Technology, closure and heterogenous engineering; the case of the Portuguese Expansion," in *The Social Construction of Technological*

Systems: New Directions in the Sociology and History of Technology, Bijker, W., Pinch, T. and Hughes, T. P. (eds.), The MIT Press, Cambridge.

Law, J. (1986) "Power/knowledge and the dissolution of the sociology of knowledge," In *Power, Action and Belief: A New Sociology*, Law, J. (ed.), Routledge, London.

Law, J. (1988) "The anatomy of socio-technical struggle," in *Technology and Social Process*, Elliot, B. (ed.), Edinburgh University Press, Edinburgh.

Law, J. (1991) "Introduction" in *A Sociology of Monsters: Essays on Power, Technology and Domination*, Law, J. (ed.), Routledge, London.

Mackenzie, D. (1991) *Inventing Accuracy: A Historical Sociology of Missile Guidance*. MIT Press, Cambridge, Mass.

Mackenzie, D. and Wajcman, J. (eds.) (1985) *The Social Shaping of Technology*. Open University Press, Milton Keynes.

Marsden, P. V. and Lin, N. (eds.) (1982) *Social Structure and Network Analysis*. Sage, London.

Mitchell, J. (1969) *Social Networks in Urban Situations*. University Press, Manchaester.

Newell, S. and Clark, P. A. (1990) "The importance of extra-organisational networks in the diffusion and appropriation of new technologies," *Knowledge: Creation, Diffusion, Utilisation*, **12**, 199–212.

Nonaka and Takeuchi, (1995) *The Knowledge-Creating Company*. Oxford University Press, Oxford.

OST, (1993) *Realising our Potential, a Strategy for Science, Engineering and Technology*. HMSO, London.

OST, (1998) *Our Competitive Future: Building the Knowledge Driven Economy*. HMSO, London.

Pelz, D. C. (1983) "Quantitative case histories of urban innovations: Are there innovating stages?" *IEEE Transactions on Engineering Management*, **30**, 60–67.

Quinquennial Review (1996) "Report of the review panel and the government's response," TCS publication.

Robertson, M., Swan and Newell, S. (1996) "The role of networks in the diffusion of technological innovation," *Journal of Management Studies*, **33**(3), 333–360.

Robertson. M, Scarborough. H. and Swan J. (1997) "Knowledge, networking and innovation: A comparative study of the role of inter- and intra-organisational networks in innovation processes," a paper submitted to the *13th EGOS Colloquium, Organisational Responses to Radical Environmental Changes*, Budapest, 3–5, July.

Rogers, E. M. (1962) *Diffusion of Innovations*. Free Press, New York.

Rogers, E. M. (1983) *Diffusion of Innovations*. Free Press, New York.

Rogers, E. M. and Shoemaker, F. S. (1971) *Communication of Innovations: A Cross-Cultural Perspective*. Free Press, New York.

Saren, M. A. (1984) "A classification and review of models of the intra-firm innovation process," *R&D Management*, **14**(1), 11–24.

Schroader, R. G., Van de Ven, A. H., Scudder, G. D. and Polley, D. (1989) "The development of innovative Ideas," in *Research on the Management of Innovation*, A. H. Van de Ven *et al.* (eds.), Harper and Row, New York.

Schumpeter, J. A. (1943) *Capitalism, Socialism and Democracy*. Harper and Row, New York.

Scott, J. (1991) *Social Network Analysis: A Handbook*. Sage, London.

Senker, P. and Senker, J. (1994) "Transferring technology and expertise from universities to industry: Britain's teaching company scheme," *New Technology, Work and Employment* **9**(2), 81–92.

Slappendel, C. (1996) "Perspectives on innovation in organizations," *Organization Studies*, **17**(1), 107–129.

Steward, F. and Conway, S. (1996) "Informal networks in the origination of successful innovations," in *The Dynamics of Cooperation in Industrial Innovation*, Coombs, R., Richards, A., Saviotti, P. and Walsh V. (eds.), Edwar Elgar, Cheltenham.

Swan, J. (1996) "Managing the process of technological innovation: A cognitive perspective." *4th International Workshop on Managerial & Organizational Cognition*, Stockholm.

Teaching Company Scheme, (1997) *Grant Application and Proposal Form.*

Tichy, N., Tushman, M. and Fombrun, C. (1979) "Social network analysis for organisations," *Academy of Management Review,* **4**(4), 507–519.

Tushman, M. L. and Scanlan, T. J. (1981) "Boundary spanning individuals: Their role in information transfer and their antecedents." *Academy of Management Journal,* **24**(1), 83–98.

Van de Ven, A. H., Angle, H. L. and Poole, M. S. (1989) *Research in the Management of Innovations: The Minesota Studies.* Harper and Row, New York.

Van Poucke, W. (1980) "Network constraints on social action: Preliminaries for a network theory," *Social Networks,* **2**, 181–190.

Whipp, R. and Clark P. A. (1986) *Innovation and the Auto Industry: Product, Process and Work Organisation.* Pinter, Londer.

Williams, R. and Edge, D. (1991) "The social shaping of technology: A review of UK research concepts, findings, programmes and centres," in *Research on the Social Shaping of Technology in France, Germany, Norway, Sweden, the United Kingdom and the United States,* Dirkes, M. and Hoffman, U. (eds.), Wissenschaftszentrum Berlin fur Sozialforschung, Berlin.

Wolfe, R. A. (1994). "Organizational innovation: Review, critique and suggested research directions," *Journal of Management Studies,* **31**, 405–431.

Zaltman, G., Duncan, R. and Holbeck, J. (1973) *Innovations and Organisations.* Wiley, New York.

Chapter 8

Organisations, Networks, and Learning: A Sociological View

Reiner Grundmann*

In this chapter, I relate the concept of organisational learning to the recent literature on industrial districts and innovation networks. The network literature so far has not applied findings from organisational learning to network learning in a straightforward way. Usually, learning is equated with innovation and 'absorptive capacity'. However, the same virtues that distinguish networks from ordinary market exchange may turn into vices which stifle the innovative process. Core firms in industrial sectors seem to be particularly prone to this. Conversely, collaboration in knowledge intensive industries typically is both an entry ticket to an information network and a vehicle for rapid communication about new opportunities and obstacles. The chapter ends with a proposed outline for further research.

Keywords: Networks, innovation, organisations, learning.

Introduction

Networks have become an increasingly fashionable subject not only in the management and business literature. It is a phenomenon which is crossing disciplinary boundaries. However, despite increased

*E-mail: r.grundmann@aston.ac.uk

research in this field, we know little about the appropriateness of network forms and contextual variables (Child and Faulkner, 1998). It is an open question whether, in reality, strategic alliances and joint ventures are growing in importance or if the phenomenon is the result of academic opportunism and business hype. The lack of consensus is mainly due to the fact that empirical research on co-operation has focused on selected types of businesses and industries or on certain types of alliances. Additionally, network analysis is not a formal or unitary theory. Rather, it is a loose federation of approaches (Burt, 1980) or even a 'terminological jungle in which every newcomer may plant a tree' (Barnes, 1972; quoted in Burt, 1980:79). Twenty years ago, Granovetter (1979) observed that 'most network models are constructed in a theoretical vacuum, each on its own terms, and without reference to a broader common framework'. This was at the beginning of renewed interest in network forms of interaction within the social sciences (cf. Simmel, 1922 for an early example). Although many studies have appeared since there has been little headway in terms of unifying network theory. Furthermore, all we know about the history of scientific development suggests that we should be pessimistic about easy attempts of theoretical unification. A way forward could be to investigate specific problems which arise in the field of research.

In this chapter, I address the issue of how networks learn. Therefore, I present selected findings from the recent literature on industrial and innovation networks. Specifically, I address two key issues, first, how are organisational learning and network learning related to each other? Secondly, which variables are useful in explaining successful network innovations? Underlying this discussion is, of course, the assumption that learning capabilities and innovation are closely interlinked.

The chapter has the following structure. In the first section, I discuss the concept of learning in organisations and networks. In the following sections, I present findings from the recent literature which advance different independent variables to explain industrial innovation. These are modes of coordination and uncertainty, market position, culture and connectedness. Based on these, areas for further research are identified.

Learning in Networks and Organisations

A common feature in the recent literature is the finding that economic variables do not fully explain the emergence of networks. Examples from innovation studies, industrial policy, and policy networks identify three issues: uncertainty, interpersonal trust, and structural embeddedness (Mayntz, 1993). Uncertainty makes the calculation of risk impossible because, since each decision situation is unique, there are no known probabilities. As there are no cases which can be compared (Knight, 1921; Schon, 1982) other mechanisms have to be found to reduce uncertainty. Some have suggested that normative and cognitive actor orientations ('culture') can provide such a mechanism. Trust and face-to-face interaction are crucial in the formation of innovation networks and industrial districts (Polanyi, 1958; Saxenian, 1991; von Hippel, 1988). Structural embeddedness points to the fact that institutional economics tends to under-socialise economic action for which it compensates with an over-socialised notion of society. This is underpinned by a functionalist outlook which suggests that institutions arise as efficient solutions to economic problems (Granovetter, 1985; Pratt, 1997; for the problem of functionalism in the social sciences, see Elster, 1983). Both trust and embeddedness emphasise the cultural dimensions of industrial innovation. However, culture has an enabling and constraining influence on every social activity, including innovation.

There is an abundance of studies on regional, industrial and innovation networks which are said to have a competitive advantage in comparison to large-scale, vertically integrated firms (Best, 1990; Sabel and Zeitlin, 1985; cf. also Lazonick, 1990; Florida and Kenney, 1990). Such networks consist of firms specialised in particular tasks but flexible in serving a wide range of market demands. They produce goods for premium price market segments, employ skilled well-paid workers and use flexible, multi-purpose assets (Piore and Sabel, 1984; Sabel, 1989). Large-scale, vertically integrated firms, on the other hand, produce standardised goods at low cost by unskilled workers operating with special purpose assets (Chandler, 1977; 1990). It is a contentious issue whether either of these institutional forms can serve as panacea for crisis ridden industries or for regions trying to (re-)gain a competitive edge (Pratt, 1997). Recently, it has been proposed that networks can become dysfunctional in the same way as other institutional forms (Podolny and Page, 1998). It has been observed that companies of all sizes in a wide range of industries are experiencing considerable uncertainty caused by rapid changes taking place in their competitive and technological environments. Electronics, new materials and biotechnology are the main examples. Industry faces turbulence or even disruption, largely because individual firms cannot control these changes (Dodgson, 1993:77).

> "In a world of uncertainty, where the probabilistic calculus is ruled out, rules, norms, and institutions play a functional role in providing a basis for decision-making, expectation and belief. Without these 'rigidities', without social routine and habit to reproduce them and without institutionally conditioned conceptual frameworks, an uncertain world would present a chaos of sense data in which it would be impossible for the agent to make sensible decisions and to act" (Hodgson, 1988:205).

Although new institutionalism has addressed this problem there are at least two competing strands: economic and sociological (Hall and Taylor, 1996 add historical institutionalism). The former analyses economic gains which accrue through networking (trans-action cost efficient) and thus remains in the realm of utility calculation and goals-means relations. The latter stresses normative and cognitive dimensions which transcend economic rationality (although one might ask if it only transcends short-term rationality or rationality as such).

> "In markets, the standard strategy is to drive the hardest possible bargain in the immediate exchange. In networks, the preferred option is often one of creating indebtedness and reliance over the long haul. Each approach thus devalues the other: prosperous market traders would be viewed as petty and untrustworthy shysters in networks, while successful participants in networks who carried those practices into competitive markets would be viewed as naïve and foolish" (Powell, 1990:303).

Economic institutionalism, as advanced by Williamson (1975; 1985), argues that repeated negotiations of complex business relationships are time-inefficient and may become prohibitively expensive. This view is summarised by Romo and Schwartz (1995:880): 'In many business situations, therefore, limits on human cognition make it cost-efficient to replace open-market exchanges with inter- or intraorganisation networks composed of long-term, taken-for-granted routines' (cf. Simon, 1945). Romo and Schwartz assert that while Williamson's line of argument may explain long-term business relationships, it does not account for spatial concentration of industry especially since the availability of modern transport and communication technologies (cf. Arthur, 1990).

Picot *et al.* (1996) argue that traditionally, hierarchical mechanisms have been implemented within firms and market mechanisms between firms. However, they suppose that this simple assignment of mechanisms to organisational forms is no longer valid. Market mechanisms are increasingly used within firms and hierarchical links are often extended to inter-firm relations. Simultaneously, we observe an increase in the geographical distances over which hierarchical as well as market relations are established. Thus, the firm tends to disintegrate organisationally and physically — a development triggered and supported by new information and communication technologies.

These technologies permit the implementation of new organisational forms that have more efficient coordination and incentive structures which weaken the firm's traditional boundaries. This view was supported by a recent study of biotechnology firms which found that learning is a social process occuring at the boundaries between firms, universities, research laboratories, suppliers and customers (Powell *et al.*, 1996; cf. also Etzkowitz, 1997; Powell, 1990). Powell *et al.* also point out that inter-organisational collaborations are not merely a way to compensate for the lack of internal skills:

> "A firm's value and ability as a collaborator is related to its internal assets, but at the same time, collaboration further develops and strengthens those internal competencies. Firms deepen their ability to collaborate not just by managing relations dyadically, but by instantiating and refining routines for synergistic partnering.... The development of cooperative routines goes beyond simply learning how to maintain a large number of ties. Firms must learn how to transfer knowledge across alliances and locate themselves in those network positions that enable them to keep pace with the most promising scientific or technological developments" (Powell *et al.*, 1996:119–120).

The existing literature has paid little attention to the question of how learning in networks occurs focusing instead on organisational learning. Thus, before tackling the question of network learning, a digression into the literature on organisational learning is in order. Broadly speaking, there is generally a distinction between the cognitive and behavioural dimensions of organisational learning. Cheng and van de Ven (1996:607) give the following definition of organisational learning: 'An experiential process of acquiring knowledge about action-outcome relationships and the effects of environmental events on these relationships'. Similarly, Levitt and March (1988:320) claim that organisations learn 'by encoding inferences from history into

Table 1. Typology of learning (after Weick, 1991).

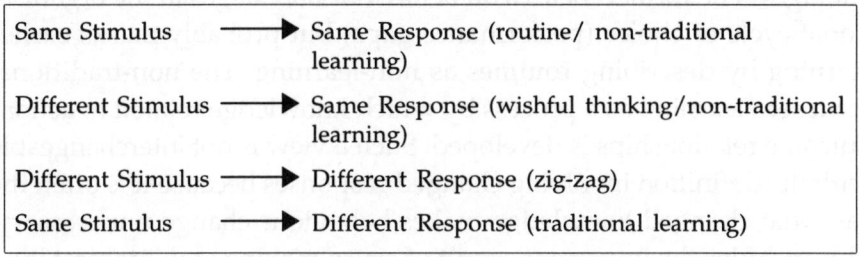

Same Stimulus	⟶	Same Response (routine/ non-traditional learning)
Different Stimulus	⟶	Same Response (wishful thinking/non-traditional learning)
Different Stimulus	⟶	Different Response (zig-zag)
Same Stimulus	⟶	Different Response (traditional learning)

routines that guide behaviour'. This definition is based on a stimulus-response model where learning occurs only in cases where the same stimulus leads to a different response from the evolving unit. Weick (1991) provides a formal distinction between behavourial ('traditional') and cognitive ('non-traditional') learning (Table 1).

Weick discusses four possible combinations which arise when combining binary stimuli and responses. In the traditional view, only the last case (same/different) counts as true learning because it depicts an endogenous behavioural change. In the non-traditional view, the focus on action outcomes is replaced with an interest in cognitive changes and two changes count as learning. Routines can be the product of cognitive changes; and sticking to the same course of action following a different environmental stimulus can also count as learning. Learning is seen as a cognitive process by means of which environmental changes are interpreted by the organisation in such a way that no altered course of action is necessary. However, routines can lead to competency traps when procedures are improved that yield limited or no competitive advantage. Relying too much on earlier success could be self-defeating since this would lead to a rigidity of standard operating procedures and to over-trust in a benevolent environment. As Hedberg (1981:14) puts it: 'Because organisations in benevolent environments often have weaker incentives to improve their performances, they may be subject to strategic stagnation'. This may lead to a decline in performance both at the level of the firm and at the level of industrial districts (see below).

Both traditional and non-traditional views have their (dis-)advantages. The former focuses on behaviour and stages in the organisational cycle of choice (performance gaps) but probably masks earlier learning by describing routines as non-learning. The non-traditional form sees learning as a process by which knowledge related to action-outcome relationships is developed. Such a view is not interchangeable with the definition involving changed responses because it is often the case that changed knowledge makes behaviour change unnecessary. This includes the important aspect of sense-making and interpretation which has become so important in the organisation literature (but it also raises the more problematic point about organisational memory which makes investigation difficult). It should be noted that in both accounts the combination of different-different is not seen as an instance of learning since they do not make rational comparisons. Every change in the environment is perceived differently and the organisation reacts in different ways.

Behavioural theories of learning were attractive because of their simplicity. The rise of cognitive learning theories can be partly explained by the fact that, in the real world, organisations sometimes cannot monitor the outcomes of decisions before they take new decisions about similar problems. On some occasions, organisations have no previous experience. Then organisations have to learn from samples of one or fewer, as March *et al.* (1991) put it. Organisations develop techniques such as scenarios or 'near-histories' to deal with this problem: 'Small pieces of evidence are used to construct a theory of history from which a variety of unrealised, but possible, additional scenarios are generated'. (March *et al.*, 1991:5). Air traffic systems or nuclear power plants provide examples of organisations that learn from near-histories or 'incidents'. If the number of events cannot be increased then it may be possible to increase the number and diversity of observers. Sometimes such techniques lead to unreliable results as in superstitious learning or the ambiguity of success. Superstitious learning occurs when the connections between actions and outcomes are ill-specified: promotion is taken to indicate high level of performance but in reality is awarded because of institutional conformity. Promoted individuals may then develop self-confidence leading them to overestimate their abilities to make sound decisions. Secondly, learning can be difficult to identify because indicators of success are redefined as are the levels of aspiration. In such circumstances, there may be political disputes centred on claims and counter-claims about success and its causes (Levitt and March, 1988; Hatch, 1997).

Argyris and Schon's (1978) distinction between single-loop and double loop learning is relevant here. Single loop learning is modelled on cybernetics where learning results from negative feedback generated by observing the consequences of action and using this knowledge to adjust subsequent action in order to modify future outcomes. Double loop learning involves reflection on the goals, strategies and routines themselves and presupposes that organisations question the basic values and mechanisms of organising. This is similar to Weick's definition of non-traditional learning. In both cases, cognitive changes within the organisation are paramount whereas behavioural changes are only related in an indirect fashion. In the organisational literature several types of learners have been proposed (Cyert and March, 1963; Weick, 1979; Coleman *et al.*, 1966; Rogers, 1962):

- Leaders: learn either by trial-and-error (i.e., from their own past experience, conceived as success or failure);[1]
- Followers: emulate the behaviour of others (i.e., 'vicarious learning', learning from the experience of others, (cf. diMaggio and Powell, 1991; Rose, 1991);
- Laggards: adopt a course of action because everyone else has done so.

---

[1]'Both success and failure contain the seeds of change. A persistent subjective sense of success leads to sense of competence and a willingness to experiment. A persistent subjective sense of failure produces instability in beliefs and disagreement among organisational participants with respect to both preferences and action' (March, Sproull and Tamuz, 1991:7).

There are two interesting things to note. First, it is not clear if leaders are 'better learners' because they might lead the market because of their sheer size. Economic power might protect large firms from failures to learn (at least for a while) no matter if learning is conceived as coming from experience or from near-history. It might be that in a specific case the laggard has experimented more than leaders and followers but still failed. So there is an ambiguity about the reference point of 'leading'. This ambiguity can be eliminated if we define leadership in relation to technological innovativeness, trial-and-error behaviour, and learning from near-history. If one applies March's (1991) influential and parsimonious distinction between *exploration* and *exploitation* in organisational learning it then seems obvious that leaders should show a good balance between exploration and exploitation of novel ideas. Followers are more likely to pursue exploitation whereas laggards are weak in exploring alternatives. This results in a stage model where the leader finds a successful response to an environmental stimulus by trial-and-error. The emulators copy this solution, and by the time the laggards adapt to the new situation, leaders have moved on to something new. The distinction between explorative and exploitative learning has the advantage that it can be applied equally well to cognitive and behavioural dimensions.

How does the above discussion relate to learning in networks? It has been claimed that networks provide learning benefits because they preserve greater diversity of search routines than hierarchies and convey richer, more complex information than markets. Here we can distinguish between formal, contractual relationships and informal relationships based on goodwill trust (Sako, 1992). In both types of networks, trustworthiness is essential: 'The most useful information is rarely that which flows down the formal chain of command in an organisation, or that which can be inferred from price signals. Rather, it is that which is obtained from someone you have dealt with in the past and found to be reliable'. (Powell, 1990:304;

cf. Podolny and Page, 1998:62; Gulati and Gargiulo, 1999). Several points can be made here:

- networks provide more diversity and redundancy than firms;
- within networks trustworthiness exists as a resource ('goodwill trust');
- reputation based on trust facilitates future cooperation.

According to Cohen and Levinthal (1989), networks increase *absorptive capacity* since R&D not only generates new information but also develops the firm's ability to exploit existing knowledge from the environment. Networks should still show an unequal distribution in terms of a clear distinction between leaders, followers and laggards. Although network terminology sometimes suggests that there is a lateral exchange between equals empirical examples demonstrate that there are asymmetric relations between heterogeneous actors (Ebers, 1999; Hoffmann and Scherr, 1999; Lunnan and Kvalshagen, 1999). Before proceeding to the next section it is worthwhile reiterating the main points so far: the important but potentially pernicious role of routines; the significance of cognitive learning from 'near-history'; double loop learning; formal and goodwill trust; and asymmetric relations between network partners.

Industrial Districts: Uncertainty and Coordination

Robertson and Langlois (1995) advanced a helpful typology to conceptualise networks. The authors operate with the (plausible) assumption that no institutional form is superior to any other. Rather, it is the type of uncertainty faced by firms that make one institutional design perform better than another. Extending Coase's one-dimensional market-firm spectrum they analyse networks along twon dimensions: ownership and coordination. Both dimensions express different aspects of integration and can be scaled accordingly. This gives four pure types (and two mixed):

Table 2. Network typology according to Robertson and Langlois (1995).

Degree of		Degree of coordination integration	
		low	high
Ownership	low	Marshallian district District	'Third Italian'
Integration		Venture Capital network	
			Jap. Kaisha network
	high	Holding company	Chandlerian Firm

The Chandlerian divisional firm (formal vertical integration) and holding companies are both 'pure types' that are highly integrated through ownership but are different with regards to coordination where the vertically integrated firm scores highly. Conversely, a loose network structure (having a low degree of ownership integration and coordination integration) is exemplified by the Marshallian industrial district. Examples from 19th century Britain include small firms in Lancashire and the Midlands. Firms in industrial districts have high degrees of specialisation, rely heavily on market mechanisms for exchange and concentrate on a single function in the production chain. The major advantages 'arise from simple propinquity of firms, which allows easier recruitment of skilled labor and rapid exchanges of commercial and technical information through informal channels' (Robertson and Langlois, 1995:548–549).

Compared to the classical industrial district, industrial districts in Northern Italy ('Third Italy') show higher degrees of coordination. Their cooperation extends to such activities as business services, provision of capital (establishment of cooperative banks), sponsorships of trade fairs as well as domestic and international marketing ventures. 'As a result, small firms are able to sell their output in world markets and to gain some of the benefits of scale economies which continue to compete strongly with each other' (Robertson and Langlois,

1995:549). Apart from these ideal types, the authors present two mixed categories: venture capital networks and Japanese kaisha networks. The first is exemplified by the Silicon Valley high-technology industrial network (Saxenian, 1990; 1991) and the second by Japanese car manufactures. Silicon Valley and Route 128 provide a variation of the Marshallian industrial district as some coordination is provided by venture capitalists. As with the Third Italy, there has been organic growth of the regional network although it took more than two decades for the nuclei of small firms to develop into large clusters. However, the organic character of the network is somewhat reduced by the venture capitalists since they superimpose themselves on the network of producers. Thus, they create a stronger degree of ownership integration typical of industrial districts, although falling short of Chandlerian vertical integration. In the case of Japanese kaisha networks, there are long-term relational contracts between car producers and suppliers which results in a high score on the coordination dimension (Florida and Kenney, 1991). Car manufacturers, however, often also own a substantial stake in their suppliers thus increasing the degree of coordination.

The purpose of Robertson and Langlois' exercise is to show that there is no single degree of integration, that suits all purposes. Vertical integration is effective in cases where the quick and cheap appropriation of new capabilities is an issue. 'This may indeed help explain the prevalence of large vertically integrated companies in the periods that Chandler chronicles' (Robertson and Langlois, 1995:554). At other times, vertical integration can inhibit innovation. In order to develop this argument, Robertson and Langlois distinguish three different forms of uncertainty under which firms operate: parametric, strategic and structural. Under parametric uncertainty, firms do not know the exact design of cloth or tile that will be demanded next season but do know what is needed to produce cloth and tiles. Industrial districts are effective in dealing with this type of uncertainty. Strategic uncertainty means that capabilities have to be re-arranged more radically but still within known boundaries. Here, vertically integrated firms are likely to have the advantage over market-based networks. Under conditions of structural uncertainty, firms do not know the relevant parameters of future products and decentralised innovative networks are likely to have an advantage. At the early stages of a product life cycle structural uncertainty is likely to be highest but this reduces as the product progresses to maturity. Higher degrees of co-ordination are of little help since this increases certainty merely by reducing the sources of reliable information (Robertson and Langlois, 1995:557). However, in situations where there are many innovative ideas and the form of the user product is in flux it is important to retain as many options as possible. Robertson and Langlois introduce the notion of 'innovative network' or 'network of networks' to capture the idea that these links need to capture both the widest range of information while keeping down collection and processing costs.

What Drives Innovation? The Benefits of Constraint

Raider (1998) studied the effects of networks and market competition on innovative activity. Reviewing literature related to the Schumpeterian hypothesis that large, monopolistic or oligopolistic firms are the driving forces of capitalist innovation she states that 'empirical examinations of the relationship between market concentration and innovation have yielded little in the way of conclusive results' (Raider, 1998:3; cf. also Cohen and Levin, 1989). Raider uses Burt's structural hole theory to overcome deficiencies of economic approaches to innovation and especially to inter-industry differences. According to Burt (1993), autonomous actors who are able to bridge a 'structural hole' occupy a favourable position in the social structure by connecting other actors who are themselves not connected. 'This brokerage or gatekeeping location in the social structure is a position of competitive advantage because it offers the opportunity to access diverse information, to control the transfer of information between disconnected parties, and to identify and broker transactions between otherwise disconnected parties' (Raider, 1998:5). In contrast, actors who are tied to a few densely connected actors are constrained because they lack the information benefits of accessing diverse social and economic worlds and have few if any brokerage opportunities. Raider applies this idea of information benefits to R&D activities: 'By transacting with many markets, industries access diverse information not only about terms of trade but about technological competencies, know-how, and technological needs of diverse buyers and suppliers. Because structurally autonomous industries connect otherwise disconnected markets, they are uniquely positioned to observe and act on these technological opportunities' (Raider, 1998:7).

In order to test her hypothesis that such firms are able to cross-fertilise innovation processes, Raider develops two indicators of innovative activity: R&D intensity and rate of innovation. The first indicator is a measure of research input and the latter measures innovation outcomes (see Raider, 1988:8–9). It is proposed that firms in positions of structural autonomy invest more in R&D and introduce innovations at a higher rate. Based on US-data reported by the Federal Trade Commission in 1976 and the Yale Survey on R&D (Levin *et al.*, 1987), Raider (1998:13) finds that 'concentration alone fails to significantly predict either R&D intensity or rate of innovation. Instead, technological opportunity has the most significant effect: industry R&D intensity is greater for industries with relatively new infrastructure (plant, property, and equipment), and where users and government laboratories and agencies are important sources of knowledge'. Nor did Raider confirm the hypotheses that structural autonomy facilitates innovation. Instead, innovative activity was found to be higher when firms were faced with higher constraints (Raider, 1998:12; cf. Hedberg's point that organizations in benevolent environments have weaker incentives to improve performance). Hence, innovative activity is highest when competition is most severe: 'Concentrated industries facing extreme downstream and upstream competitive contexts devote a greater proportion of their resources to research and development and experience higher rates of innovation, suggesting adversity is a motivator for innovative activity' (Raider, 1998:19).

Raider does not discuss the possibility that firms which are structurally autonomous in network structures may be constrained by internal factors whereas firms which are said to be structurally constrained may be able to chose their strategy more autonomously. This seems to be plausible because structurally autonomous firms are usually large (and old) firms while structurally constrained firms are usually small (and new) firms. Dodgson (1993:85) makes a similar point: 'New, small firms are unconstrained by many of the factors identified by the organisation theory and strategic management litera-ture which may limit large firms' freedom to respond quickly. These factors include the development of inhibitory loops and introspection and conservatism in learning, and the firm-specific nature of strategic technology management related to historical circumstances and the cumulative and tacit nature of technological knowledge that constrain novel activities'. In this interpretation, there are two different types of constraint: one stemming from the network position and the other from organisational characteristics. Two other significant findings from Raider's study pertain to network size and density. Network size is strongly negatively associated with R&D activity (which suggests that exposure to more markets discourages innovation) while network density has a strong positive association with R&D intensity. 'This seems to imply that small, densely interconnected networks facilitate innovative activity' (Raider, 1998:16). I will return to this point below.

The Decline of Industrial Districts

The starting point for much recent work in interorganisational network research was that transaction cost economics failed to explain the concentration of industrial districts or the propensity of firms to innovate or join networks. Instead, norms, values, tacit knowledge, trust and face-to-face interaction were seen as playing a key role in the formation of innovation networks and industrial districts (Polanyi, 1958; Saxenian, 1991). However, too strong an adherence to norms can lead to institutional rigidities which undermine flexibility. Two examples, one from Germany and one from the United States illustrate the point.

Grabher (1993) argues that industrial districts will decline if they begin to lose their learning capabilities. Drawing on the concept of double-loop learning (Argyris and Schon, 1978) he argues that uncommitted and unspecific resources are essential and serve unforeseeable uses. Apart from personnel, material and technological, resources also include the ability to self-question. Grabher rightly points to the fact that norm-following behaviour as espoused by neo-institutionalism, not only has the advantage that it reduces uncertainty but also the disadvantage that it leads to rigidities in business routines which can be detrimental in the long run. Inspired by Granovetter (1982), he analyses the decline of the Ruhr area after the second World War which suffered from a threefold lock-in: functional, cognitive and political. Long-term stability and predictability of demand for iron and steel helped create and maintain close links between core and supplier firms. The latter largely gave up their own R&D aimed at developing new products for potential new customers. Personal ties between management of supplier and core firms supplanted the suppliers' own marketing strategy. As for the cognitive lock-in, the Ruhr area became a homogenous regional culture which produced a common world-view precluding competing perceptions of reality and 'groupthink' emerged (Morgan, 1986:91; cf. March's 'competence trap'). Finally, there was a close co-operative relationship between industry and the regional administration of Nordrhein Westphalia which went unchallenged for decades. This insular political culture was created by conservative Social Democrats, trade union leaders and patriarchal industrialists. The power base of the largest Social Democratic party within Germany resided within the coal, iron and steel industry. Following a severe crisis in the 1980s, over 60,000 jobs were lost, plants closed down and began to shift their economic activities to the South of Germany (although wage rates were similar). Firms in the Ruhr no longer saw themselves as coal, iron, and steel companies but redefined themselves as 'technology-based firms'. Mannesmann, for example, has become one of the major players in the telecommunications market. Plant engineering, environmental technology, mechanical engineering and electronics also took centre stage in the region's strategic reorientation.

A complementary analysis has been made by Romo and Schwartz who examined the process of de-industrialisation in New York State as a consequence of industrial migration. They tested the 'universally held assumption' of standard comparative cost models which explain industrial migration above all by high wages and union power (see Hayes and Wheelwright, 1979; Harvey, 1982; Storper and Walker, 1989). Their data from 1960 to 1985 show that 70% of migrating firms relocated to regions where prevailing sector-specific wages were as high or higher than those in New York. Like Grabher, they use a 'structural embeddedness' approach to analyse New York State as an industrial complex in which spatial proximity led actors to develop a local language with a tacit dimension which required the transacting parties to be in long-term physical proximity. As they point out, this argument is not an extension of transaction cost analysis since proximity is not always more efficient for the firm (Romo and Schwartz, 1995:881). Trust can be seen as the essential ingredient in making regional innovation networks work. Spacial proximity fosters norms of trust since there is a greater probability of future interaction among neighbouring businesses and decreases the likelihood of free-riding (Axelrod, 1984). It is useful to introduce a distinction between *trust enhanced* networks and *unenhanced* networks. The former 'improves information availability, it reduces transaction costs, contributes to positive association, and may ameliorate negative externalities' (Carney, 1998:460). Unenhanced networks are created where large enterprises simply outsource parts of their business, sometimes stimulated by a search for low costs and flexibility. In so doing, they become 'hollowed out', losing the ability to innovate and design their core products (Cohen and Levinthal, 1991).

Romo and Schwartz use the concept of 'intersectoral resource dependence' (Burt, 1983) in order to understand how local economies grow up around key establishments (core firms) which are supplied by many smaller 'peripheral' establishments which differ in four important aspects:

- core establishments tend to buy from many local suppliers, peripheral plants tend to sell their output to a small number of local customers (i.e., core establishments);
- core establishments tend to be large, peripheral plants tend to be small;
- core establishments tend to have a multidivisional structure, peripheral plants tend to be unitary organisations;
- core establishments tend to be part of large multilocal or multinational companies, peripheral plants tend to be part of smaller, unilocal firms.

The upshot of their argument is that core establishments are responsive to cost incentives while peripheral plants are not. 'While migrating core establishments overwhelmingly selected distant locations when New York's business climate was poor, migrating periphery plants remained in the region, even in the most unfavourable circumstances' (Romo and Schwartz, 1995:901). Two things should be mentioned, first, migrations of core establishments accounted for less than 15% of migrating plants and for 25% of job losses in the region. Second, Romo and Schwartz convincingly argue that these migrations were a response to the crisis not the cause. This raises the possibility that industrial districts have not lost their innovative capacity.

The overall finding of Romo's and Schwartz's study is that core establishments react to lost market share by moving to low cost regions. Since core firms tend to be large and old they are less capable or willing to innovate. However, comparative costs enter the picture only when the production culture begins to erode: 'A weakened production culture cannot respond quickly to competitive challenges, and this may lead management of core firms to choose the spatial fix instead of innovating' (Romo and Schwartz, 1995:903). Although using different language, both Grabher and Romo and Schwartz reach similar conclusions regarding the cause of industrial decline. What Grabher labels the petrifying of a regional culture Romo and Schwartz call the erosion of culture. The latter explain decline through their core-periphery model in which core firms become unable to break with managerial habits that are outmoded. In a similar vein, Grabher points to a lock-in process that undermines innovative capacity. In both cases, culture is no longer seen as enabling but as constraining. In order to pursue this question further I will examine innovation networks in the next section.

Networks of Learning: Innovation Networks

I am now in a position to return to the question of links between informal goodwill-trust and network learning. Powell *et al.* (1996) studied the fields of electronics and biotechnology where they argue that radical new developments restructure mature industries or, in Schumpeter's term, lead to 'gales of creative destruction'. The locus of innovation is likely to be found in networks provided that knowledge is distributed widely and that it brings competitive advantages: 'To stay current in a rapidly moving field requires that an organization have a hand in the research process. Passive recipients of new knowledge are less likely to appreciate its value or to be able to respond rapidly' (Powell *et al.*, 1996:119). Collaborations in high-tech industries typically reflect more than just a formal contractual exchange: 'When the first author presented the chief executive officer (CEO) of Centocor with a list of his firm's formal agreements, he observed that it was 'the tip of the iceberg' — it excludes dozens of handshake deals and informal collaborations, as well as probably hundreds of collaborations by our company's scientists with colleagues elsewhere. Beneath most formal ties, then, lies a sea of informal relations' (Powell *et al.*, 1996:120). Cooperation within R&D networks is both an entry ticket to an information network and a vehicle for the rapid communication about new opportunities and obstacles. However, innovative activities cannot be reduced to a simple process of information acquisition. They emerge from informal R&D collaboration that usually takes on a more formal and contractual character once such projects lead to feasible products. Generally speaking, involvement in cooperative R&D projects widens the horizon of a firm's personnel and makes it sensitive to new developments and projects which are 'out there' or could be initiated by the firm.

Technological breakthroughs tend to level the playing field for involved firms as they also reap rewards by exploiting new opportunities created by the networking process. Within R&D networks, satisfying mutual needs becomes more important than the desire to defeat opponents. Although, the structural position of a firm within the network is decisive. The authors formulate the hypothesis that 'the greater a firm's centrality in a network of relations at any given time, the greater the number of subsequent R&D collaborations, controlling for prior collaborative R&D activity' (Powell *et al.*, 1996:122). While small firms need the financial support of larger firms, larger companies mainly want to get access to the research expertise of smaller firms.

Based on their study, Powell *et al.* (1996) conclude that age is unimportant and size is an outcome rather than a determinant of network activities. As a result of reciprocal network learning the boundaries between firms become more permeable. In contrast to earlier work that established a liability-to-newness-hypothesis (Hannan and Freeman, 1989; Stinchcombe, 1965) the authors observe a liability-to-unconnectedness in biotechnology and other fields with similar characteristics. Firms prefer sustaining their ability to learn instead of appropriating new capabilities: exploring instead of exploiting. This leads to interdependence rather than independence, to network embeddedness rather than vertical integration. Actors in these fields realise that 'knowledge is sophisticated and widely dispersed and not easily produced or captured inside the boundaries of a firm...nor readily available for purchase' (Powell *et al.*, 1996:143). Firms do not collaborate merely to acquire resources and skills they cannot produce internally. Participation in networks cannot be understood according to a make-or-buy logic. For example, in Silicon Valley, employees in different firms tend to share information with trusted friends paying little attention to the sacrosanct boundaries of formal organisation (Saxenian, 1990). This raises important questions about trade secrets and intellectual property rights. Property rights for technological information in emerging fields are hard to establish because of problems in counting, trading or contracting transactions. There are also severe limitations in patent law and property rights protection. As Zeckhauser (1996:12747) puts it: 'Our patent system was developed for the era of the better mouse trap and its predominantly physical products, whereas today we are designing better mice'. Therefore, much technological information is not patented and what is patented has fuzzy demarcations. New organisational forms are a promising approach to contracting difficulties for technological information: 'Webs of relationships, formal and informal, involving universities, start-up firms, corporate giants, and venture capitalists play a major role in facilitating the production and spread of technological information' (Zeckhauser, 1996:12743).

Because technological information is easily accessible, on the one hand all major players know what their competitors are planning or what feasible options they have. Mansfield (1985) observed that information concerning the details of new products and processes generally leak out within one year. On the other hand, the virtual availability of technological information presupposes that firms are engaged in R&D and keep abreast of new developments in the field. The properties of technological information (hard to count and trade) and the unreliability of policies for the protection of intellectual property lead to new organisational forms ranging from traditional mergers to networks involving universities, start-up firms, corporate giants and venture capitalists. Interestingly, Zeckhauser then goes on to make the case for regions in which social capital and trust are institutionalized, hinting explicitly at Silicon Valley and Route 128 (see Carney, 1998).

Zeckhauser's findings are supported by research undertaken by Tidd and Trewhella (1997) who argue that in the absence of strong property rights, firms prefer to develop 'difficult-to-codify' technologies in-house. Such 'tacit technologies' are seen to provide a more durable source of competitive advantage than those which can easily be codified. This is based on the work of Nonaka and Takeuchi (1995) who identify the transformation of tacit to explicit knowledge as the critical link in organisational learning. Since all knowledge originates with the individual, organisations can only learn through communication, face-to-face interaction and 'knowledge networks' (Nonaka and Takeuchi, 1995). The latter span overlapping organisational boundaries, tap and accumulate new knowledge then share it throughout the firm.

Conway (1997) describes the role of gatekeepers (Allen, 1977; Cohen and Levinthal, 1989; von Hippel, 1987) in the innovation process. He distinguishes between the roles of *liaisons* (the inter-mediary between two or more groups), *bridges* (links the gatekeepers of two groups) and *link-pins* (actors with overlapping membership of two or more groups) who all link the firm to the environment. The case studies of successful innovating firms show that these gatekeepers experience a tension between their organisational colleagues and the outside world which they have to mediate. Additionally, they have to link into a diverse range of external cliques: 'The trick to being in the right place at the right time, is to be in a lot of places' (Conway, 1997:231). Because gatekeeping is an individual rather than an organisational role this makes the innovating firm vulnerable since the 'absorptive capacity' depends on personal networks of small numbers of employees. This raises the problem of personal allegiance because if a gatekeeper switches to another firm the original employer loses not only the employee but also the skills and knowledge which might be difficult to replace. Even so, firms always have to face the risk of informal arrangements in which representatives are not controllable at each and every step, a phenomenon well known in principal-agent theory (Pratt and Zeckhauser, 1985). In such a case, adherence to professional networks might turn out to be stronger than adherence to the firm. Firms must exercise trust when entering knowledge intensive networks because they run the risk of being exploited by the co-operating partners or of losing precious assets. Firms must balance the risks associated with being part of a network with the risks of being excluded. In order to reduce uncertainty, the rules and norms associated with institutions offer a basis for decision-making. This opens the way for fruitful co-operation but also leads to the danger of exploitation, over-embeddedness and stagnation. This paradox is sometimes resolved by transforming the network into an organisation through merger and take-over. Sometimes the tension is increased because of the need to experiment and co-operate and the need to reduce uncertainty by introducing routines produce more and more tension.

Further Research

Neither the network nor the organisational learning literatures share a great deal of consensus about appropriate theories, methods or data. Furthermore, there seems to be a real danger of sample bias in that theories only pick 'appropriate' cases which support their point. Ideally, this predicament could be overcome by studying the same data ('paradigm cases') with different theories or by testing one theory against various empirical examples (see Hassard, 1993). Both of these strategies would require the large-scale co-operation of researchers, a possibility that may emerge in the near future if the trend towards networking is real. I briefly outline a third possibility which, at least, tries to take into account the danger of sample bias when applying theories to cases and yet could be carried out by a relatively small research team. To repeat, as a point of departure we should be aware of the likelihood that certain theories are better with respect to specific empirical problems. In order to contravene such a potential flaw in research design, and to increase the robustness of findings, a strategy should be chosen which is cognisant of the fact that specific fields of investigation might represent special cases or lend themselves to an all-too-easy theoretical interpretation. Thus, the following questions for further research seem to follow from this literature review which has emphasised two different types of learning (exploration and exploitation) and the importance of network structure. Table 3 tries to capture some of the empirical examples encountered in this chapter, along the dimensions of learning and network centrality. The columns distinguish between learning capabilities where exploration stands for firms which are good at learning, either by conducting trial-and-error, double loop learning or learning from near-history. Exploitation denotes emulating or adapting behaviour. The rows distinguish between core or periphery firms.

Industrial districts should be investigated more closely to establish whether core firms are declining in their innovative capacity or if

Table 3. A two-dimensional typology of learning networks.

	Exploration	Exploitation
Core	Terza Italia, Silicon Valley (1)	Postwar Ruhr Area, ('unenhanced networks') (2)
Periphery	Knowledge intensive start-ups Technological opportunities (3)	Suppliers of (2) (4)

there are counter examples. If so (maybe *La terza Italia*, Silicon Valley), the successful cases should be carefully compared with the unsuccessful cases to explain the differences. Some writers have emphasised the beneficial role of industrial districts (Sabel *et al.*) whereas others have stressed issues of lock-in processes and decline (Romo and Schwartz *et al.*). Furthermore, comparisons need to be longitudinal since today's successful examples may be tomorrow's failures.

Co-operation in knowledge-intensive sectors needs to be investigated to establish whether it represents a special case of networking or if it reveals something generalisable. The hypotheses would concern the role of informal agreements and trust, of professional associations and of technological opportunities (according to Raider's finding that R&D intensity is greater for industries with relatively new infrastructure). As a result of these tentative reflections, it is possible to set out the following independent variables which could influence successful co-operation: age, size, learning type, tech-nological opportunities, position in network (centre/periphery) and relations in network (formal/informal). As various contributions have indicated, *age* and *size* of a firm do not have a direct impact on innovative activity (*pace* Schumpeter, Stinchcombe, Hannan and Freeman). Instead, network theorists point to the liability of unconnectedness. Secondly, network position does not have a direct impact on innovative activity as innovative firms can be found at the periphery and the centre of a network as well as being structurally autonomous or constrained. Thirdly, technological opportunities tend to have a positive impact on innovative activity. Fourthly, informal, trust-based relations enhance co-operation. Thus, this literature review suggests only two positive propositions, the first related to technological opportunities and the second concerning informal relationships based on goodwill-trust. To operationalise suitable variables it is necessary to distinguish between learning within the organisation and learning in the context of networks.

In the first instance, organisational structures and practices should be examined in order to find out if and how interpretations about near-histories are carried out within the organisation. In the second, we look at the set of actors coming together in network settings where each actor is a representative of an organisation (gatekeeper). Learning in this context is likely to follow a different path and logic than the learning process in their parent organisations. People in such settings are involved in projects which go beyond the routines of their parent organisations. Therefore, the last two propositions would read as follows: Organisations with cognitive learning resources are more likely to be innovative than those without. Since they are more likely to co-operate with others, a related point pertains to the networking activities. Here, one can put forward the hypothesis that the greater the diversity of external cliques a gatekeeper links into, the higher the chance to be in the right place at the right time. (See Fig. 1 for a graphical representation of the variable set.)

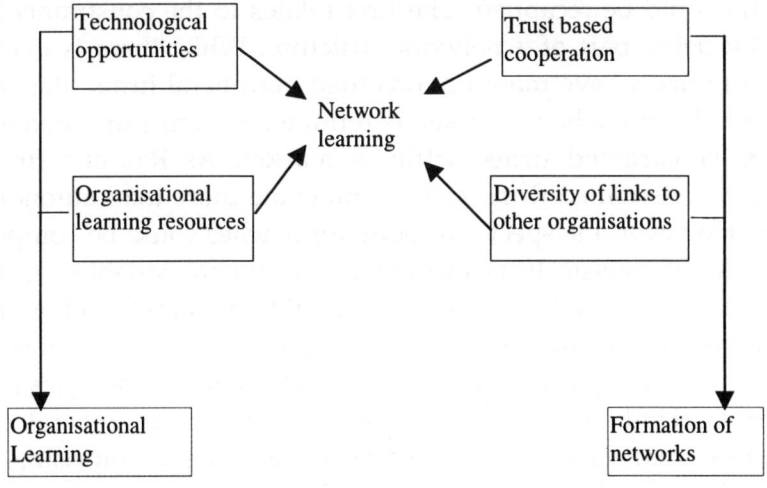

Fig. 1. Variables of network learning.

The graph distinguishes between the intra-organisational dimension in the left-hand column and the inter-organisational dimension in the right-hand column. Propitious intra-organisational conditions enhance organisational learning while propitious inter-organisational conditions enhance the formation of networks. Only if both come together can network learning occur. Successful innovative activity is a result of technical opportunities, organisational learning resources, trust-based cooperation and diversity of links. While these conditions appear to be necessary it is far from clear if they are sufficient for network learning.

Conclusion

If there is a common theme in the recent literature on industrial districts and innovation networks it is the attempt to leave transaction cost approaches behind and to develop the structural embeddedness argument from different perspectives. There are three main points which should be recapped. The first relates to the constraints firms face by being part of a network structure. While there is evidence that core firms have more options than peripheral firms this seems not to hold true when the issue is reframed in terms of autonomous versus constrained firms within a market. As Raider's findings suggest, constrained firms seem to innovate more than autonomous firms. However, I suspect that both approaches could be compatible upon closer analysis. If we include organisational variables, as Romo and Schwartz do in their analysis, it might be that internal organisational constraints are sometimes stronger than network constraints (Raider does not pay attention to this aspect since her analysis is only about market-based networks). This raises a more general point: organisational variables and network variables (comprising both constraints and incentives) have rarely been examined together — which clearly calls for more research. From the existing literature it is not clear if specific characteristics of industrial sectors (knowledge intensity in biotech and electronics) explain institutional structures or vice versa. In other words, it could be that the knowledge intensity in high-tech sectors makes certain institutional forms obsolete whereas they persist in less knowledge intense sectors.

Secondly, the literature on industrial districts and on knowledge based networks contradict both the Schumpeterian thesis and the liability-of-newness argument. The same virtues that distinguish networks from ordinary market exchange may turn into vices which stifle the innovative process. Routines, norms and cultural embeddedness can have an inhibiting influence on innovating activities. While culture can have detrimental effects on the inclination to innovate, recent research on knowledge based industries emphasises that age is not decisive for innovation. Rather than newness, it is unconnectedness that becomes a liability for firms. Boundaries between firms become more and more permeable, knowledge exchange and transfer happens more quickly than in traditional sectors of industry. This is an interesting, and at first sight, counter-intuitive finding. If true, it suggests that classical vertical firms are entirely different from innovation networks when facing issues of trade secrets. It seems that vertical firms protect only a fraction of knowledge compared to the amount of knowledge circulating in innovation networks. It is here where we see a degree of openness and cooperation between firms which is unprecedented. This is not to say that problems of intellectual property do not exist — quite to the contrary. But they are so difficult to solve by traditional means (patents and the like) that new organisational forms are increasingly sought as an answer.

The third point relates to the problem-solving capacities of networks. While there is an optimistic view that recommends networks as a new institutional panacea for industrial reorientation, others are more careful and even pessimistic. The decline of industrial districts points to the liability of political and cultural over embeddedness, part of which is the lack of organisational learning resources. Podolny and Page (1998) have summarised the sociologists' interest in 'the prevalence and functionality of organisational forms that could not be classified as markets or hierarchies. As a result of this work, we now know that network forms of organisation foster learning, repre-sent a mechanism for the attainment of status or legitimacy, provide a variety of economic benefits, facilitate the management of resource dependencies, and provide considerable autonomy for employees. However, as sociologists move away from critiquing what are now somewhat outdated economic views they need to balance the exclusive focus on prevalence and functionality with attention to constraint and dysfunctionality'. This seems to be an apt reminder that *all* institutional forms have their drawbacks when it comes to enhance innovative activity. Further research should focus on systematic comparisons between successful and unsuccessful core firms in industrial districts and enhance our understanding of knowledge intensive-networks.

Acknowledgements

A previous version of the argument was presented at the 15th EGOS Colloquium, Warwick, July 4–6 1999, Subtheme 2: 'Interorganizational relations and networks: Methodologies and theories'. I would like to thank the participants of this workshop, and my colleagues Steve Conway and Ossie Jones for their constructive comments.

References

Argyris, C. and Schon, D. A. (1978) *Organizational Learning*. Addison-Wesley, Reading, MA.

Arthur, W. B. (1990) "Positive feedbacks in the economy," *Scientific American*, 92–99.

Axelrod, R. (1984) *The Evolution of Cooperation*. Basic Books, New York.

Best, M. H. (1990) *The New Industrial Competition: Institutions of Industrial Restructuring*. Harvard University Press, Cambridge, MA.

Burt, R. S. (1980) "Models of network structure," *Annual Review of Sociology*, 6, 79–141.

Burt, R. S. (1983) *Corporate Profits and Cooptation: Networks of Market Constraints and Directorate Ties in the American Economy*. Academic Press, New York.

Burt, R. S. (1992) *Structural Holes*. Harvard University Press, Cambridge, MA.

Carney, M. (1988) "The competitiveness of networked production: The role of trust and asset specificity," *Journal of Management Studies*, 35, 457–479.

Chandler, A. D. (1977) *The Visible Hand*. Harvard University Press, Harvard, MA.

Chandler, A. D. (1990) *Scale and Scope: The Dynamics of Industrial Capitalism*. Harvard University Press, Cambridge, MA.

Cheng, Y.-T. and Van De Ven, A. (1996) "Learning the innovation journey: Order out of chaos?" *Organization Science*, 7, 593–614.

Child, J. and Faulkner, D. (1998) *Strategies of Cooperation. Managing Alliances, Networks, and Joint Ventures*. Oxford University Press, Oxford, New York.

Cohen, M. D., March, J. and Olsen, J. P. (1972) "A garbage can model of organisational choice." *Administrative Science Quarterly*, 17, 1–19.

Cohen, W. M. and Levin, R. C. (1989) "Empirical studies of innovation and market structure," in *Handbook of Industrial Organization*, R. Schmalensee and R. D. Willig (eds.), Amsterdam: North Holland, 1059–1107.

Cohen, W. M. and Levinthal, D. A. (1989) "Innovation and learning: The two faces of R&D," *The Economic Journal*, 99, 569–596.

Coleman, J. S., Katz, E. and Menzel, H. (1966) *Medical Innovation. A Diffusion Study*. Bobbs-Merrill, Indianapolis.

Conway, S. (1997) "Strategic personal links in successful innovation: Link-pins, bridges and liaisons," *Creativity and Innovation Management*, 6, 226–233.

Cyert, R. M. and March, J. G. (1963) *A Behaviorial Theory of the Firm*. Prentice-Hall, Englewood Cliffs, NJ.

DiMaggio, P. and Powell, W. (1983) "The iron cage revisted: Instituional isomorphism and collective rationality in organizational fields," *American Sociological Review*, **48**, 147–160.

Dodgson, M. (1993) "Learning, trust and technological collaboration," *Human Relations*, **46**, 77–95.

Ebers, M. (1999) "How to control a partner successfully: A relational analysis of inter-organizational governance structures in the German construction industry," paper presented at the *15th EGOS Colloquium*, Warwick, 4–6 July.

Elster, J. (1983) *Explaining Technical Change*. Cambridge University Press, Cambridge.

Etzkowitz, H. (1997) "From zero-sum to value-added strategies: The emergence of knowledge-based industrial policy in the states of the United States." *Policy Studies Journal*, **25**, 412–424.

Florida, R. and Kenney, M. (1990) *The Breakthrough Illusion: Corporate America's Failure to Move from Innovation to Mass Production*. Basic, New York.

Florida, R. and Kenney, M. (1991) "Transplanted organizations: The transfer of Japanese industrial organization to the US," *American Sociological Review*, **56**, 381–398.

Grabher, G. (1993) "The weakness of strong ties: The lock-in of regional development in the Ruhr area," *The Embedded Firm*, G. Grabher (ed.), Routledge, London, 255–277.

Granovetter, M. (1979) "The theory-gap in social network analysis," in *Perspectives on Social Network Research*, P. W. Holland (ed.), Academic, New York, 501–518.

Granovetter, M. (1982) "The strength of weak ties. A network theory revisted," in *Social Structure and Network Analysis*, P. Marsden and N. Lin (eds.), Sage, London, 105–130.

Granovetter, M. (1985) "Economic action and social structure: The problem of embeddedness," *American Journal of Sociology*, **91**, 481–510.

Gulati, R. and Gargiulo, M. (1999) "Where do interorganizational networks come from?" *American Journal of Sociology*, **104**, 1439–1493.

Hall, P. and Taylor, R. (1996) "Political science and the three new institutionalisms," *Political Studies*, **44**, 936–957.

Hannan, M. T. and Freeman, J. (1989) *Organizational Ecology*. Harvard University Press, Cambridge, MA.

Hassard, J. (1997) *Sociology and Organization Theory: Positivism, Paradigms and Postmodernity*. Cambridge University Press, Cambridge.

Hatch, M. J. (1997) *Organization Theory. Modern, Symbolic and Postmodern Perspectives*. Oxford University Press, Oxford.

Harvey, D. (1982) *Limits to Capital*. Chicago University Press, Chicago.

Hayes, R. J. and Wheelwright, S. C. (1979) "The dynamics of product life cycles," *Harvard Business Review*, **57**, 127–136.

Hedberg, B. (1981) "How organizations learn and unlearn," in *Handbook of Organizational Design*, P. C. Nystrom and W. H. Starbuck (eds.), Oxford University Press, Oxford, 3–27.

Hodgson, G. (1988) *Economics and Institutions*. Cambridge University Press, Cambridge.

Hoffmann, W. and Scherr, M. (199) "The configuration of strategic alliances: A contingency approach to the analysis and configuration of inter-company collabroative relationships," paper presented at the *15th EGOS Colloquium*, Warwick, 4–6 July.

Knight, F. (1921) *Risk, Uncertainty and Profit*. Houghton Mifflin, New York.

Lazonick, W. (1990) *Business Organization and the Myth of the Market Economy*. Cambridge University Press, Cambridge.

Levin, R. C., Klevorick, A. K., Nelson, R. and Winter, S. G. (1987) "Appropriating returns from industrial research and development," *Brookings Papers on Economic Activity*, **3**, 783–820.

Levinthal, D. A. (1991) "Organizational adaptation and environmental selection — interrelated process of change," *Organizational Science*, **2**, 140–145.

Levitt, B. and March, J. G. (1988) "Organizational learning," in *Annual Review of Sociology*, **14**, W. R. Scott and J. Blake (eds.), Annual Review Inc, Palo Alto, CA, 319–340.

Lunnan, R. and Kvalshagen, R. (1999) "Aquiring knowledge in alliances: The impact of individual characteristics," paper presented at the *15th EGOS Colloquium*, Warwick, 4–6 July.

Mansfield, E. (1985) "How rapidly does new industrial technology leak out?" *The Journal of Industrial Economics*, **34**, 217–223.

March, J. G. (1991) "Exploration and exploitation in organizational learning," *Organization Science*, **2**, 71–87.

March, J. G. and Olsen, J. P. (1976) *Ambiguity and Choice in Organizations*. Universitetsforlaget, Bergen.

March, J. G., Sproull. L. S. and Tamuz, M. (1991) "Learning from samples of one or fewer," *Organization Science*, **2**, 1–13.

Mayntz, R. (1993) Policy-Netzwerke und die Logik von Verhandlungs-systemen. in A. Héritier (ed.), *Policy-Analyse. Kritik und Neuorientierung. Politische Vierteljahresschrift, Sonderheft*, **24**, 39–56.

Morgan, G. (1986) *Images of Organization*. Sage, Beverly Hills.

Nonaka, I. and Takeuchi, H. (1995) *The Knowledge-Creating Company*. Oxford University Press, Oxford.

Picot, A., Ripperger, T. and Wolff, B. (1996) "The fading boundaries of the firm: The role of information and communication technology," *Journal of Institutional and Theoretical Economics-Zeitschrift für die Gesamte Staatswissenschaft*, **152**, 65–79.

Piore, M. J. and Sabel, C. F. (1984) *The Second Industrial Divide: Possibilties for Prosperity*. Basic Books, New York.

Podolny, J. M. and Page, K. L. (1998) "Network forms of organization," *Annual Review of Sociology*, **24**, 57–76.

Polanyi, M. (1958) *Personal Knowledge*. Routledge and Kegan Paul, London.

Powell, W. W. (1990) "Neither market nor hierarchy: Network forms of organization," *Research in Organizational Behaviour*, **12**, 295–336.

Powell, W. W., Koput, K. and Smith-Doerr, L. (1996) "Interorganizational collaboration and the locus of innovation: Networks of learning in biotechnology," *Administrative Science Quarterly*, **41**, 116–145.

Pratt, A., (1997) "The emerging shape and form of innovation networks and institutions," in *Innovation, Networks and Learning Regions?*, J. Simmie (ed.), Jessica Kingsley Publishers and Regional Studies Association, London, 124–136.

Pratt, J. W. and Zeckhauser, R. J. (eds.) (1985) *Principals and Agents: The Structure of Business*. Harvard Business School Press, Boston, MA.

Raider, H. J. (1998) "Market structure and innovation," *Social Science Research*, **27**, 1–21.

Robertson, P. L. and Langlois, R. (1995) "Innovation, networks, and vertical integration," *Research Policy*, **24**, 543–562.

Rogers, E. M. (1962) *Diffusion of Innovations*. The Free Press, New York.

Romo, F. P. and Schwartz, M. (1995) "The structural embeddedness of business decisions: The migration of manufacturing plants in New York State, 1960 to 1985," *American Sociological Review*, **60**, 874–907.

Rose, R. (1991) "What is lesson-drawing?" *Journal of Public Policy*, **11**, 3–30.

Sabel, C. F. (1989) "Flexible specialisation and the re-emergence of regional economies," in *Reversing Industrial Decline? Industrial Structure and Policy in Britain and her Competitors*, P. Hirst and J. Zeitlin (eds.), Berg, Oxford, 17–70.

Sabel, C. F. and Zeitlin, J. (1985) "Historical alternatives to mass production: Politics, markets, and technology in nineteenth-century industrialization," *Past and Present*, **108**, 133–176.

Sako, M. (1992) *Prices, Quality, and Trust. Inter-Firm Relations in Britain and Japan*. Cambridge University Press, Cambridge.

Saxenian, A. (1990) "Regional networks and the resurgence of Silicon Valley," *California Management Review*, 89–112.

Saxenian, A. (1991) "The origins and dynamics of production networks in Silicon Valley," *Research Policy*, **20**, 423–437.

Schon, D. A. (1982) "The fear of innovation," in *Science in context*, S. B. Barnes and D. O. Edge (eds.), Open University Press, London, 290–302.

Schumpeter, J. A. (1942) *Capitalism, Socialism, Democracy*. 6th edition, Unwin, London.

Simmel, G. [1922] (1955) *Conflict and the Web of Group Affiliations*. Trans. Reinhard Bendix. Glencoe, Il.

Simon, H. A. (1945) *Administrative Behavior*. Macmillan, New York.

Stinchcombe, A. (1965) "Social structure and organizations," in *Handbook of Organizations*, J. G. March (ed.), Rand Nelly, Chicago, 142–193.

Storper, M. and Walker, R. (1989) *The Capitalist Imperative: Territory, Technology, and Industrial Growth*. Basil Blackwell, New York.

Tidd, J. and Trewhella, M. J. (1997) "Organizational and technological antecedents for knowledge acquisition and learning," *R&D Management*, **27**, 359–375.

von Hippel, E. (1997) Cooperation between rivals: Informal know-how trading," *Research Policy*, **16**, 291–302.

Weick, K. E. (1979) *The Social Psychology of Organising*. 2nd ed., Addison-Wesley, Reading, MA.

Weick, K. E. (1991) "The nontraditional quality of organizational learning," *Organizational Science*, **2**, 116–123.

Wiesenthal, H. (1995) Konventionelles und unkonventionelles Organizations-lernen," *Zeitschrift für Soziologie*, **24**, 137–155.

Williamson, O. E. (1975) *Markets and Hierarchies*. Free Press, New York.

Williamson, O. E. (1985) *The Economic Institutions of Capitalism*. Free Press, New York.

Zeckhauser, R. (1996) "The challenge of contracting for technological information," *Proceedings of the National Academy of Sciences of the United States of America*, **93**, 12743–12748.

Chapter 9

The Innovative Capacity of Voluntary and Non-Profit Organisations: Networks and the External Environment*

Stephen P. Osborne[†]

Introduction

The role of voluntary and non-profit organisations (VNPOs) in providing public services in the UK and the rest of Europe is changing. Traditionally, they have taken on a role either of providing those public services which the state could not provide or of extending the coverage and/or quality of those which it did (Mellor, 1985). However, since the early 1980s, this role has been profoundly transformed. The key factors involved in this transformation are complex and include, over the 1980s, both an ideological 'sea-change'

*This paper is based upon research funded by the Joseph Rowntree Foundation. The views expressed in it, however, are solely the responsibilitty of the author and do not necessarily represent those of the Foundation.
[†]E-mail: s.p.osborne@aston.ac.uk

in public perceptions about the appropriate role of the state in providing public services and growing concern over the spiralling cost of such services. These factors have been explored elsewhere (Mischra, 1984; Ascher, 1988; Lane, 1994) and will not be pursued here.

For VNPOs in the UK, this has led to profound changes in the societal environment within which they operate. Two issues have been especially significant. The first has been the changing profile of their income, away from voluntary sources and toward government as the major funder of their work (Osborne and Hems, 1995). The second has been the determination of the previous Conservative government to use VNPOs as part of a strategy intent upon disman-tling the direct service provision role of local government in favour of a plural system of public service provision to local communities (Rao, 1991). Whilst the present Labour government in the UK has retreated from the most negative views of local government contained in this strategy, it nonetheless sees a significant role for VNPOs in providing public services, because of their perceived potential both to contribute to the lowering of public spending, by 'levering in' additional resources to the public sector and to contribute to social inclusion across the UK (Labour Party, 1997; Osborne and Ross, 1999).

Both Conservative and Labour governments have predicated this key role for VNPOs upon two beliefs. First, the importance of competition in the provision of public services as a tool through which to gain value for money, or 'best value' in its current formula-tion. The impact of this belief, often called the 'new public manage-ment', has been analysed elsewhere (for example, Ferlie *et al.*, 1997). The second belief, less well-analysed, asserted that voluntary organisations had a special contribution to make to innovation in public services (Home Office 1990; DETR 1998). As Knapp *et al.* (1990) have noted, the innovative capacity of VNPOs has achieved 'legendary' status. However, as other work by this author (Osborne, 1998b) demonstrates, this status is based upon a demonstrable innovative role for such organisations in the nineteenth century (Webb and Webb, 1911).

This present chapter is derived from a major programme of research on VNPO innovation in the context of personal social services in the UK. The study combined an empirical mapping of innovative activity by voluntary organisations in this field together with an exploration of the causal factors for this activity (Osborne, 1998b). In this chapter, the focus is on the role played by networks in releasing this capacity and the impact of the external environment (see Poulton, 1988).

The chapter commences with a brief outline of the theoretical background to role of networks in the provision of public services and the innovative capacity of VNPOs. Following a discussion of the research methodology, the findings are used to explore the role of environmental networks in releasing and sustaining the innovative capacity of VNPOs.

Theoretical Background

The Concept of Networks in Organisation Theory

The concept of 'networks' has a substantive place in organisation theory which stretches back over the last three decades. This history comprises two broad traditions. First, networks have been used primarily as a metaphor through which to describe the myriad of inter-organisational relationships (for example, Collins, 1974). Secondly, there has been the use of statistical and quantitative approaches to generate complex mathematical models of inter-organisational networks (Breiger, 1976; Knokke and Kuklinski, 1984). Recent work by the editors of this volume has attempted to bridge this divide by both mapping organizational networks and developing substantive theory (Conway, 1997; Jones, 1997).

This chapter explores the use of the concept in relation to the provision of public services by VNPOs. It seeks both to map the differing networks of innovative and non-innovative VNPOs and to highlight their impact upon the management of the innovation process. In approaching this task, it is necessary to be clear about what is meant by the term *network*. Networks do not have an independent existence. Rather it is a conceptual term which refers to patterns of relationships which arise between interacting organisations (Rogers, 1987). Such a definition highlights issues of *boundary spanning* (how organisations co-operate across organisational boundaries) and *boundary maintenance* (how organisations maintain their organisational integrity in the face of such co-operation). The network literature draws on significant work in organisation studies which examines links between the external environment and organisational behaviour (Benson, 1975; Miles and Snow, 1978):

> "...to understand the behaviour of an organisation you must understand the context of that behaviour.... Organisations are bound up with the conditions of their environment.... (They) engage in activities which have as their logical conclusion adaptation to the environment" (Pfeffer and Salancik, 1978:1).

This approach was subsequently developed by Granovetter (1985) who emphasised that the environment both constrains and enables behaviour in organisations. More recently, some researchers have questioned the impact of the external environment upon the innovative capacity of organisations (de Bresson and Amesse, 1991). At a macro-level, this work has focused on the relationship between the demands of national market economies and organisational innovation (Gomulka, 1990; Nelson, 1993). Increasingly, authors such as Best (1990) and Kreiner and Schultz (1993) have argued that collaboration rather than competition has, in some cases, become the prime determinant of innovative capacity. These macro-studies have led other writers to explore the micro-level relationships between the external environment and innovation (Camagni, 1991; Alter and Hage, 1993). This present chapter draws upon this latter literature for inspiration. It will argue that the network approach to organisational behaviour is relatively new in the field of public policy and management. It seeks to explore what insights can be gained into the role of VNPOs in public and social policy implementation by utilisation of this perspective.

Networks and the Provision of Public Services

Compared to the wider organisation studies literature, use of the 'network' concept to understand public policy-making remains 'a minority interest' (Rhodes, 1997). Although, recently, there has been growing interest in links between networks and policy formulation (Rhodes, 1990; Kickert, 1993; Provan and Milward, 1995; Lynn, 1996). Kickert *et al.* (1997:1) provide a good summary of the core components of this approach:

> "The concept "policy network" connects public policies with their strategic and institutional context: the network of public, semi-public and private actors participating in certain policy fields. The concept is new in the sense that it combines insights from policy science, which focuses on the analysis of public policy processes, with ideas from political science and organisation theory about the distribution of power and dependencies, organisational features, and inter-organisational relations. As an empirical phenomenon policy networks can be found in almost every policy area."

Kickert *et al.* (1997) have developed a theory of *public governance* as an alternative to Anglo-American dominance of the *new public management* (Hood, 1991; Ferlie *et al.*, 1997). Whilst the latter stress the need for a businesslike and competitive approach to government, the former emphasise the significance of network management. The work of Kickert and colleagues at Erasmus University (Rotterdam) has contributed to the development of network management at both a theoretical level (Klijn, 1996; Kickert and Koppenjan, 1997) and at an applied level through the application of 'game theory' (Klijn *et al.*, 1995; Klijn and Tiesman, 1997; de Bruijn and ten Heuvelhof, 1997). This literature makes a significant contribution to our understanding of the role and functioning of networks in policy arenas. A key limitation is that, with the exception of O'Toole *et al.* (1997), studies in this area concentrate on policy *formulation* rather than the *implementation* process. This means that key actors in the planning, provision and management of public services are not included in the network. With regard to VNPOs, this is particularly important given the emphasis laid on 'networking' as an approach to organisational activity (Osborne and Tricker, 1994; Osborne, 1999). Nonetheless, little attention has been paid to using the concept of 'networks' as a way to understand the role and management of such organisations.

Innovation, Social Policy and the Innovative Capacity of VNPOs

Deutsch (1985:19) argued that the study of innovation and development of associated substantive theory have become the 'major potential modifiers of social and political theory' in the late twentieth century. This view has special resonance in the study of contemporary social policy in the UK where service innovation has of late become a significant tenet of such policy. The aspiration and expectation was that policy reforms aimed at stimulating competition (Porter, 1985) in public services would promote user responsiveness and cost efficiency (Osborne and Gaebler, 1992). This approach is particularly evident in the reform of the personal social services in the UK over the last decade. The Kings Fund Institute (1987), for example, argued for the centrality of innovation within community care policies whilst the National Institute for Social Work argued that innovation should be 'almost synonymous' with good social work practice (Smale and Tuson, 1990). Such concerns also made themselves increasingly felt within the policy formulation process. A good example was the Griffiths Report (Griffiths, 1988) on community care and the subsequent government White Paper (Department of Health 1989). Legislative reform highlighted the failure of community care policies over the previous 20 years and sought to create a new framework for their improvement. Central to this framework was the concept of the 'enabling state' as the planner and funder of such services but at the same time:

> "...making the maximum possible use of voluntary and private sector bodies to widen consumer choice, *stimulate innovation* and encourage efficiency" (Griffiths, 1988, para. 1.3.4) [my emphasis].

This emphasis was reinforced in the subsequent White Paper with its emphasis on the development of services 'which meet individual needs in a more flexible and innovative way' (Department of Health, 1989:para. 3.4.3). Wistow *et al.* (1994; 1996) have argued that this concentration on plural service provision and innovation marked a paradigmatic shift rather than incremental change in terms of social policy formulation and the provision of personal social services. A cornerstone of this paradigmatic shift was a belief in the innovative capacity of VNPOs. However, as Knapp *et al.* (1990:199) have noted, this attributed capacity has invariably been based on conjecture rather than sound empirical facts:

> "It is rarely the case that one can detect either an empirical or a conceptual basis for (it).... The pioneering characteristic of voluntary organisations has been cited so frequently as to become legendary. But like all the best legends the truth has sometimes been colourfully embellished to make a better story."

The roots of this legendary capacity lie in the pioneering role such organisations took in the development of nineteenth century public services. In the UK, documentary evidence compiled by Webb and Webb (1911) was subsequently transformed into a cornerstone of public and social policy-making. Beveridge (1948:302) asserted that the voluntary sector needed 'to pioneer ahead of the state and to make experiments' whilst the Ministry of Health (1959) declared that innovation was likely to be the most valuable contribution made by voluntary organisations to social welfare services in Britain. Such assertions have continued to the present day (Wolfenden, 1978; Wagner, 1988; Home Office, 1990; Commission on the Future of the Voluntary Sector, 1996). In the research literature there has also been some support for the inherent innovative capacity of VNPOs (Peyton, 1989) while others have argued that innovation is limited to small VNPOs (Johnson, 1987) or that it is actually a disguised form of organisational development rather than true innovation (Kramer, 1981).

The research on which this chapter is based explored the roots and mapped the extent of innovation in VNPOs with specific focus is on the role of networks. As such the research goes beyond the policy formulation of Kickert and his colleagues to explore the role of networks in the strategic and operational management of the agencies involved in the provision of public services.

Research Methodology

The research focused on Personal Social Services (PSS) in the UK which are aimed at providing social support services both for children in need and their families and for adults with special needs (see Osborne, 1996). The research had three components. First, a concern with structuring the organisational field within which VNPO innovative activity took place. The survey established both the types of *innovations* produced by VNPOs and the organisational characteristics of the *innovators*. This, for the first time, provided an empirical account of innovative activity in VNPOs and an estimate of the extent of   such activity. The   second part of   the   research was concerned with the causality and processes associated with innovative activity. Three cross-sectional case studies provided the opportunity to test four possible explanations for the capacity of VNPOs to innovate. These explanations derived from the existing literature were as follows:

- that their innovative capacity was a function of the structural characteristics of these organisations — *the organisational explanation* (Mellor, 1985);
- that their innovative capacity was a function of their internal environment — *the cultural explanation* (Johnson, 1987);
- that their innovative capacity was a function of their relationship to their external environment — *the environmental explanation* (Poulton, 1988);
- that their innovative capacity was a function of the institutional framework of their activities — *the institutional explanation* (Singh *et al.*, 1991).

The third part of the study integrated findings from parts one and two to build a preliminary theoretical model to explain the innovative capacity of VNPOs. By collecting in-depth data, the research captured the richness and diversity of different localities and also explored the relationships between governmental bodies and the local political economy. Three localities were chosen to represent a rural (*Southshire*), a suburban (*Bellebury*) and an urban social environment (*Midwell*).[1] Southshire is a large rural county in the south of England recognised as having high levels of rural

---

[1] In order to preserve the confidentiality of the respondents to this study, fictitious names have been used throughout.

deprivation and isolation. Bellebury, also in the South of England, is an affluent suburban area with large elderly population. Finally, Midwell is an inner-city locality in the West Midlands which scores highly on a range of indicators of social deprivation.

Defining and Classifying Innovation

A key stage in the development of a methodology for this study was the creation of a conceptual definition and an empirical classification of innovation within the wider field of organisational change. Details of this task are reported in Osborne (1998a) and the key points are summarised here.

An extensive literature review carried out as part of the exploratory work for this research uncovered a range of definitions of innovation in the organisation studies literature and drew out four key themes. First, *innovation represents newness to the organisations concerned* — though this often implied new to the organisation itself rather than in an absolute sense (Kimberly, 1981). Secondly, *invention and innovation are different;* invention is the creation of new knowledge whilst innovation is the application of that knowledge (Aiken and Hague, 1974; Freeman, 1982). Thirdly, *innovation is both a process and an outcome.* Finally, and perhaps most significantly, *innovation involves discontinuous change,* as opposed to gradual organisational development where the emphasis is upon continuity (Herbig, 1991). This discontinuity produces the *transformation* of organisational capabilities (Urabe, 1988) which is the crucial feature of innovation compared to other forms of change (Tushman and Anderson, 1985). Innovation breaks with the existing production paradigm in the same way that railways replaced, rather than improved, canals as a form of transport. As discussed below, this does not necessarily make innovation normatively superior to organisational development although it does present different managerial challenges.

The literature review revealed five common approaches to the classification of innovation. Four of these, though useful in illuminating some aspect of innovation, were too linear and simple to capture the true complexity of innovation. For example, see the approach to *product* and *process* innovation described by Bessant and Grunt (1985). The approach favoured here was taken from the work of Abernathy *et al.* (1983) who, as part of a major re-analysis of US economic development, created a typology which classified innovation in terms of its impact on both the market *and* the production processes of an organisation. This approach did not set-up unnecessary dichotomies but allowed issues of continuity and discontinuity to be explored in the context of interrelationships between production process and market. Consequently, this author was able to produce a typology of organisational change in the human services within which innovative activity was differentiated from developmental activity (Osborne, 1998a; Fig. 1). The key point is that this typology allows an exploration of relationship between the impact of a change both upon the beneficiary group of a VNPO and upon the services that it provides. This approach produces four types of organisational change:

➢ *total* (where change is both new to the organisation, or may even be the creation of a new organisation itself, and serves a new beneficiary group — such as the 'buddy system' devised by several VNPOs for supporting people with AIDS),

➢ *expansionary* (where change involves offering an existing service to a new beneficiary group — such as the UK probation service utilising methods devised for supporting juvenile offenders to work with young offenders),

➢ *evolutionary* (where change involves providing a new service to an existing beneficiary group — such as the implementation by local authorities of the 'care management' approach to community care services for elderly people in the UK), and

➢ *developmental* (where an existing service to an existing beneficiary group are modified or improved — such as refining the role of home care assistants by agencies supporting the elderly).

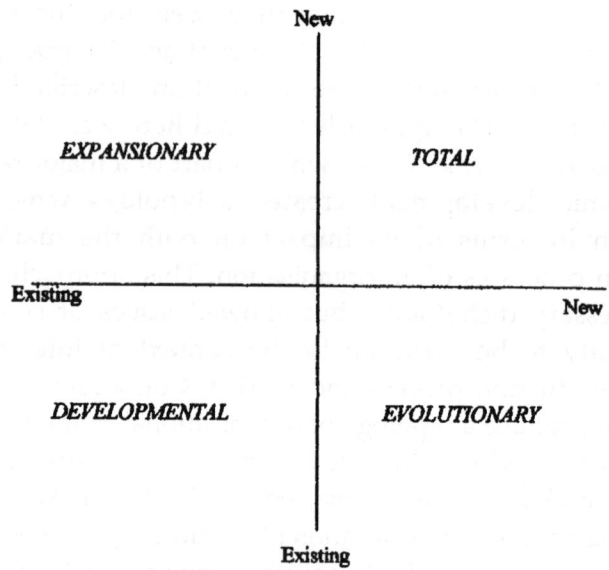

Key: 'x' axis - services of the agency
'y' axis - needs of the client [end-user] group

Fig. 1. A typology of organisational change in human services.

A particularly important characteristic of this typology is that it allows the differentiation of innovative activity (comprising the first three types of organisational change specified above) from developmental activity (the latter type of organisational change). These two activities were denoted respectively *innovative* and *developmental* activity. The third type of activity explored in this study was *traditional* activity in which an organisation continued to provide the same services to the same beneficiary group without any form of change.

The Survey

The first stage of the research utilised a postal survey to establish what the key actors within VNPOs understood by *innovation* and to discover the extent of actual innovative activity. The survey asked respondents if their organisation had initiated an innovation over the past three years and, if so, to describe it. This gave important insights into how the managers of VNPOs themselves perceived innovation and allowed the researcher to classify each innovation according to the typology of organisational change described above (providing an 'objective' perspective on the extent of this capacity). Finally respondents were asked a series of questions about the basic characteristics of their organisation such as size/nature of staff, age and main source of organisational funding.

The population of VNPOs in each locality was established through local intermediary organisations. In total, 376 organisations were surveyed across the three localities, and 196 replied which gave a response rate of 52.1%. However the real response rate was actually higher because of organisational morbidity amongst registered VNPOs: some cease to exist although continuing to be listed in voluntary or charitable registers (Osborne and Hems, 1995).

The Case Studies

Twenty four organisations were involved in total divided equally across three different localities. These organisations were categorised according to the innovation typology described above (innovative, developmental and traditional). Innovative VNPOs were those with activities falling within the total, expansionary and evolutionary segments of the typology; developmental organisations were those falling within the developmental segment; and the traditional organisations were those with no new or changing organisational activity. The VNPOs concerned included self-help and support groups, service providers advice and information centres and local intermediary bodies in the field of the personal social services.

The main sources in each case were the organisational leaders (both paid and unpaid) because of their strategic overview of the key issues and relationships involved. Where appropriate, discussions were also held with organisational staff, users and members, and key external stake-holders such as local government and other funders of their activity. The case studied involved a range of research methods including attitudinal tests, semi-structured interviews, unstructured interviews and documentary evidence. The case studies were thus strong on both data and methodological triangulation. Issues of reliability and validity for both the survey and the case studies are discussed further in Osborne (1996b).

The Impact of Networks on the Innovative Capacity of VNPOs

The remainder of this chapter examines the impact of the external environment, and the role of networks, on the innovative capacity of VNPOs. First, it is necessary to explore the strategic approach of VNPOs to their environments and their positioning within their inter-organisational fields. It will demonstrate the importance of the environment as a core driver of innovative capacity rather than any inherent characteristics of the VNPOs. Networks were used as a tool with which to describe these environmental relationships. Inter-organisational networks were also a key factor in organisational processes which facilitated the release of innovative capacity within VNPOs.

The Strategic Approach of VNPOs to their Environment

Miles and Snow (1978; see also Astley and Van de Ven, 1983; Zahria and Pearce, 1990) in a seminal work analysed the extent to which organisations have choice in the way that they interact with their environments. Reflecting the 'strategic choice' (Child, 1972) approach to organisational analysis, they argued that organisational fields, or industries, do not act monolithically; rather each organisation in that field seeks its own destiny. They proposed that organisations can decide what they want within a specific environment and can be proactive in interacting with the environment to achieve their goals. Organisational fields do not act monolithically because each organisation views its environment in relation to its own needs and goals. Miles and Snow subsequently developed four *gestalts* by which to classify the specific strategic adaptation of an organisation to its environment. These gestalts were based on the assumption that all organisations need to resolve three broad problems:

➢ the *entrepreneurial* problem, the choice of product/market domain,
➢ the *engineering* problem, the choice of technology to produce its product/service, and
➢ the *administrative* problem, the choice of organisational structures and processes to best reduce uncertainty in relation to the market and to provide optimal congruence with the environment.

The gestalts themselves were:

➢ *defender*, an organisation which produces a limited product service line in a stable environment with the emphasis on efficiency,
➢ *prospector*, an organisation which produces a broad and/or dynamic product/service line in a changing environment which avoids commitment to one specific technology,
➢ *analyzer*, an organisation which combines the above two approaches by competing both in stable markets, with a limited product/service line, and in changing markets, with a variable product/service line and in doing so matches its technology to a varied environment, and
➢ *reactor*, an organisation which makes inconsistent choices without any clear strategic direction.

The original concern of Miles and Snow was to explore the impact of these gestalts upon internal organisational adaptation. However, subsequent work (Beekum and Ginn, 1993) has expanded the approach to cover the shaping of external relationships, arguing that

Table 1. The Miles and Snow gestalts and their pattern in this study.

Gestalt	Key Features	Types of Organisation		
		Innovative	*Developmental*	*Traditional*
Defender	Limited product/service line, with an emphasis on stability and efficiency	0	2	7
Prospector	Broad/changing product and service line with a dynamic approach to its environment	4	0	0
Analyzer	Has a standard range of products/services, but also searches for new ones	5	3	1
Reactor	Makes inconsistent choices; a "non-strategy"	0	2	0

for any organisation 'its inter-organisational pattern should reflect its strategy'. Their approach, followed in this research, also expanded the typology to explore human services as well as manufacturing and technical activities. The gestalts and overall pattern of relationships found in this research are summarised in Table 1. The differing strategic approaches of the innovative and developmental VNPOs, compared to the traditional ones, is striking. The latter were almost entirely committed to the *defensive* gestalt of commitment to the 'status quo' of their services with a premium upon efficiency and a deep suspicion of external attempts to introduce change.

Table 2. Direction of service change pattern.

Types of Organisation	Decreasing the overall range of services	Maintaining the overall range of services (including substitution)	Increasing the overall range of services
Innovative	0	2	7
Developmental	0	4	4
Traditional	0	7	1

The innovative organisations were positive and proactive in their strategic approach with both the *prospector* and *analyser* gestalts prevailing amongst these organisations. In some cases, this led to an embracing of a dynamic approach to their environment as a whole whilst in others it was a case of maintaining a core of standard services but with a willingness to explore alternative models of service delivery. The developmental organisations showed no distinct overall pattern but presented a mixture of responses to their social environment. This pattern was further confirmed by the responses of these organisations to a question about their overall service pattern (Table 2). Whilst the traditional organisations were largely committed to maintaining their existing level of services (traditional *defender* activity) there was a commitment to increasing their range of services from almost all of the innovative organisations (typical of the *prospector* and the *analyser* gestalts). The difference in approaches was graphically illustrated in short passages from two of the open-ended interviews:

"We provide transport here — it is what we do. Some other (organisations) have tried to get us to change, to say that people need different things now, but its what we do" (driver for a traditional VNPO).

"Networking is very important for us. It is the way that we find out what is going on and what's needed. How else could we do it?" (Organiser of an innovative VNPO)

Thus, whilst the non-innovative organisations saw change as undermining their existing competencies and services, the innovative VNPOs took a far more proactive role in seeking ways to develop their organisations in a changing environment. To paraphrase a common management saying: innovatory VNPOs viewed change as an opportunity whilst the more traditional ones saw it as a threat.

The Inter-Organisational Field

VNPOs in this study were concerned with the provision of personal social services. The environmental complexity of the case study organisations was evaluated by discussion with organisational leaders about the key inter-organisational relationships which they needed to maintain in order to achieve their 'mission-critical' goals. *Simple environments* were defined as those where an organisation had a minimal need to interact (such as simply taking telephone referrals from other organisations for volunteer drivers, for example). *Medium environments* were those where organisations needed to interact at a significant level with one other organisation in order to achieve their 'mission-critical' goals. Finally, *complex environments* were those where organisations needed to interact with at least two, and often more, organisations in order to achieve these goals.

The analysis showed significantly different patterns in organisational environments (Table 3). The innovative organisations inhabited far more complex environments than the traditional ones while, once again, there was no clear pattern for the developmental organisations. The difference in perspectives was graphically illustrated by two respondents. When discussing external contacts a member of one traditional organisation dismissed the importance of working with other organisations in this wider environment: 'No, we don't work

Table 3. Environmental complexity of case study organisations.

Type of Organisation	Complexity			
	Single	*Medium*	*Complex*	*Total*
Innovative	0	4	5	9
Developmental	1	4	2	7
Non-innovative	6	2	0	8
Total	7	10	7	24

Table 4. Organisational linkages to their environments.

Type of Organisation	Types of Linkages			
	Isolation	*Direct*	*Network*	*Total*
Innovative	0	4	5	9
Developmental	0	4	3	7
Traditional	5	3	0	8
Total	5	11	8	24

with other organisations — no other groups offer what we do'. Conversely, the organiser of one innovative organisation saw such relationships as essential to their work: 'I used to work with these people (as a teacher). I know them and they can talk to me about their needs. I also know the people in the statutory agencies. We work together'.

A similar picture emerged when the respondents described the relationship of their organisation to its wider environment (Table 4). Again, three alternative patterns were identified: *isolation,* where there was a minimal linkage between organisation and environment; *direct,* where the linkage was direct from organization to its wider environment; and *network,* in which linkages were complex and involved the conscious negotiation of inter-organisational relationships. The pattern of these relationships confirmed those from

(i) Organisation 17

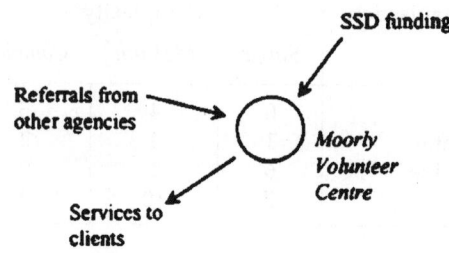

(ii) Organisation 22

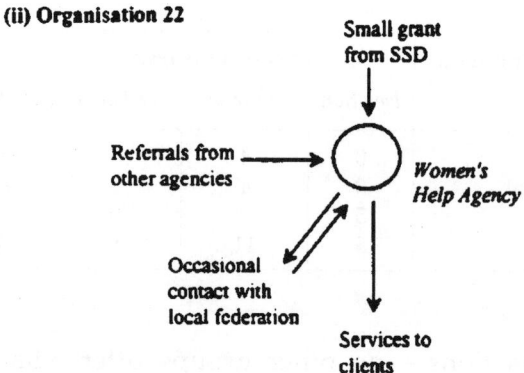

(iii) Organisation 24

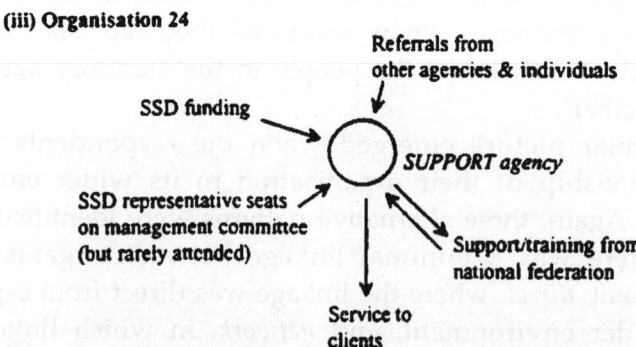

Fig. 2. Examples of network patterns of traditional organisations.

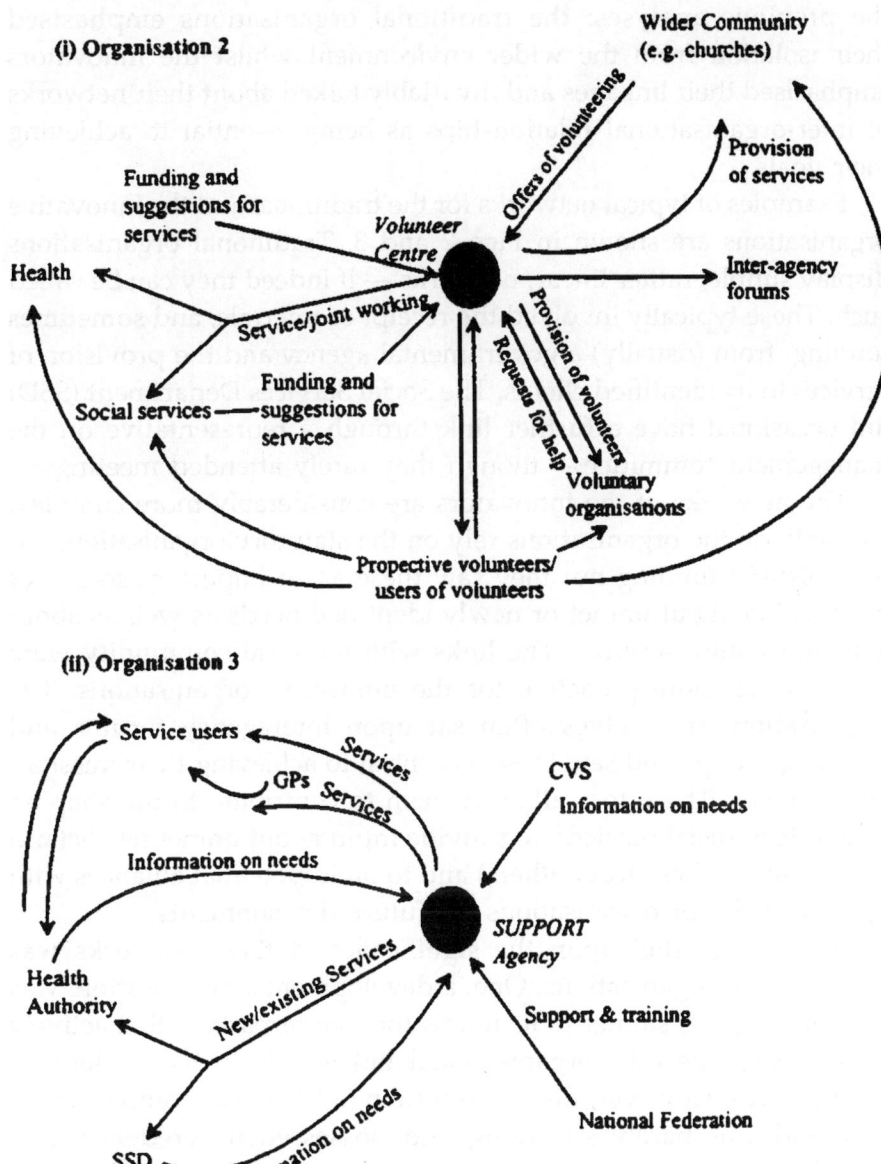

Fig. 3. Examples of networks patterns of innovative organisations.

the previous analyses: the traditional organisations emphasised their isolation from the wider environment whilst the innovators emphasised their linkages and invariably talked about their networks of inter-organisational relationships as being essential to achieving their goals.

Examples of typical networks for the traditional and the innovative organisations are shown in Figs. 2 and 3. Traditional organisations display simple, rather linear, networks — if indeed they can be called such. These typically involved the receipt of referrals, and sometimes funding, from (usually) a governmental agency and the provision of services to its identified clients. The Social Services Department (SSD) did occasional have a further link through a representative on the management committee — though they rarely attended meetings.

The networks of the innovators are considerably more complex. Not only do the organisations rely on the statutory organisations for referrals and funding but they saw these as an important source of information about unmet or newly identified needs as well as about gaps in existing services. The links with the local community were similarly far more proactive for the innovative organisations. The organisations themselves often sat upon interagency forums and planning groups and saw these as central to achieving their mission-critical goals. These fora allowed them to contribute to the shaping of statutory social services, to provide input about unmet needs (and to learn about these from others) and to build potential alliances with other agencies or organisations for future developments.

Interesting relief upon the significance of these networks was given by two organisations. One, a developmental organisation, was in many respects similar to its innovator counterparts in that actively sought to create inter-organisational linkages in order to develop new services. However, this organisation (a CVS) was comparatively new and was barely surviving and possessed no credence as a significant actor with the other organisations in its locality. This lack of network linkages severely limited its ability to fulfil its innovative potential:

"I have very little time. I need to develop more contacts
locally — they are important for my work, but all my effort
is taken up with [surviving].... Its very frustrating" (organiser
of the organisation).

"The CVS is under-resourced. It's not very effective...[it]
doesn't lack goodwill but what it does lack is the resources
and contacts to carry the words into action" (manager of a
major established local voluntary agency involved in working
with the CVS).

Confirmation of the importance of these inter-organisational relationships and networks was provided by a VNPO which provided a toy
library. This organisation was failing largely because its 'traditional'
toy library service for *children* with special needs had been established
outside existing networks of service providers and without the
legitimating support of the SSD (which went on to establish its own
similar resource). In an attempt to avoid extinction, its managerial
team tried to diversify into the provision of leisure support for *adults*
with special needs. Once again, however, it was not properly linked
into the existing service delivery network. This meant that it lacked
legitimacy with pre-existing service providers (it was seen as a child-
care organisation inappropriately trying to work with adults) and
unclear about the actually unmet needs. Consequently, it did not
have the contacts to help with the implemen-tation and maintenance
of its innovation. In a very real sense, this lack of a network of
interagency connections led to the failure of this innovation and
possible of the organisation itself.

At the descriptive level, therefore, the use of the concept of the
'network' has allowed different strategic approaches to their
environment by VNPOs to mapped and explored. It has shown
how VNPOs with different strategic gestalts adopted very different
relationships with their external environments and has demonstrated
a significant correlation between these relationships and their

innovative capacity. However, use of the 'networks' concept is not limited to exploration of the *strategic intent* of VNPOs. It is also possible to use networks as a means through which to understand the *operational management* processes of innovatory VNPOs.

Networks and the Process of Innovation

In discussing the innovation processes, respondents identified seven roles which networks played in underpinning this activity. The first was to provide *a general service context*. For example, one organisation derived its purpose from the failure of the statutory social services to provide meaningful day occupation services for adults with learning disability. It therefore began filling this gap as part of a network of service providers within the context of the overall service provision for such adults.

The second role was *to provide legitimisation for the work of the organisation*. A good example of this was one VNPO sponsored by the SSD to provide support to other VNPOs in the development of community care services. Without these links to the SSD the VNPO certainly would not have been seen as a credible organisation in its field.

The third role was *to provide sources of ideas for new service developments*. At its most basic this was simply the exchange of demographic information but more often it involved agencies working together to identify important areas of new needs or areas of unmet known needs ('service gaps'). In one case, linkages to two different statutory agencies played a key role helping a Volunteer Bureau identify areas of unmet need for adults with mental health problems and for elderly people needing practical support. Whilst both the other agencies had known of one of these needs they had never seen them in relation to each other. The position of the Bureau within its service network allowed it to bring these two needs together and to develop a service which met both needs.

The fourth role was *to facilitate the interagency planning of new services.* This could often happen through an interagency planning team co-ordinated by a VNPO. A good example was a multi-agency planning team which developed a new service for children-at-risk in the community. This team had been co-ordinated by a local child-care charity which ultimately provided the service that the team designed and resourced.

The fifth role was *to help in resource acquisition* by providing a conduit for information about funding sources and potential linkages which was a common use of networks for many VNPOs in this study.

The sixth role was *to act as a key factor in the actual implementation of an innovation.* The sexual counselling service of one VNPO was initially reliant on existing contacts with the Health Authority and GPs for disseminating information about its service and providing referrals. These were an essential precursor to the success of this innovation.

Finally, inter-organisational networks provide an important role in *the sustenance of an innovation.* (the failing CVS discussed above was a negative example or this). A more positive example was an organisation in which multi-agency commitment to a new service was the key to its survival and success.

It is important to emphasise that these networks derived their importance from being *the outcomes of other activity* rather than their creation being an activity in its own right. Such networks were the realisation of successful working relationships and derived much of their status from such activity. Where 'networking' was pursued as an end in itself it was noticeably less successful both for the organisations and innovations concerned (Osborne and Tricker, 1994).

Conclusions

This chapter has reviewed new evidence about the importance of the external environment for the innovative capacity of VNPOs. In doing so, it has found the concept of the 'network' an important tool in both the description of this role and its analysis. Important factors and relationships have been uncovered in the strategic approaches of innovative VNPOs to their environments. It is suggested that the environmental relationships of VNPOs are embedded in broader, institutional linkages which provide organisational legitimacy and help survival. This has been hypothesised previously by both Singh *et al.* (1991) and Tucker *et al.* (1992) who argued that VNPOs, in particularly those active in personal social services, are reliant on demonstrating their effectiveness to 'higher order' organisations upon which they are dependent for their main funding. However, in the personal social services, creating such links is difficult and in such circumstances innovative activity can become a surrogate for effectiveness.

Indeed it is this desire to demonstrate organisational effectiveness which might provide the common thread which binds the developmental VNPOs together. It has been found that many developmental VNPOs often claimed that their work was innovative whilst the typology used in this study demonstrated that it was not. One possible explanation for this is that the assertion of an innovative capacity was a proxy used to gain legitimacy with their key stake-holders such as local government who valued such capacity highly. Their is no definite corroboration for this hypothesis in this study and it will need further exploration. However, it does possess considerable construct validity. As was noted at the outset, 'innovation' has become prized as an highly desirable activity both within public services in general and in the personal social services in particular. It would not be suprising, therefore, if such developmental VNPOs did seek to assert heir innovative capacity as a means through which to enhance their organisational legitimacy.

Of course, innovative activity is not the only way to gain legitimacy. This study also found many examples of VNPOs providing specialist services, which were highly valued locally. Indeed, some traditional VNPOs seemingly eschewed innovation quite purposefully in order to maintain legitimacy based upon one of these other forms of organisational activity. Viewed in this context, the innovative capacity of a VNPO is indeed embedded in its environment which is itself realised through relational networks. This is because innovation performs an essential function in ensuring legitimacy by helping an organisation demonstrate congruence with its institutional environmental. This environment enables it to undertake some activities, constrains it from undertaking others or dictates how it should present its activities to gain legitimacy.

In conclusion, therefore, this paper has found the external environment of VNPOs to be a significant determinant of innovative capacity. Special importance has been attached to the strategic approach of VNPOs to their environment. The role played by 'networks' both as an analytic concept with which to understand strategy and as an operational tactic by which to manage and sustain the innovation process has also been explored. In doing this, it has been possible to describe the range of inhabit networks that VNPOs and to model their impact on innovatory activity. Such an approach is important, for two reasons. First, it has contributed to the generation of substantive theory about the innovative capacity of VNPOs and allowed the development of a model to reflect this capacity (Osborne, 1998b). The work of Kickert and his colleagues discussed above has provided a core rationale both for the existence of networks within the policy-making and implementation process. The larger research study upon which this chapter is based has developed a model which shows how individual organisations, in this case VNPOs, are located within these networks. The central component of this model has been the way in which institutional forces both constrain and enable the actions of such organisations. In this chapter, this model has been used to demonstrate how such inter-organisational networks are a significant factor in the development and release of the innovative capacity of VNPOs. Secondly, it has also provided insights into the management of such networks within plural patterns of the provision of public services. Such patterns are becoming globalised (Salamon,

1995) and pose particular problems for their management both by government and by the other actors (Osborne, 1998c; Osborne and Ross, 1999). The use of the 'network' approach offers important lessons for the practical governance of such plural service provision.

Inevitably, though, no paper can provide the definitive answer to its central preoccupation and this is the case here. In particular, two issues have been raised but not resolved. First, as noted above, are questions about the significance of the institutional context of a VNPO in constructing its innovative capacity. The second point which has not been addressed here, is the need to develop a more analytic understanding of the processes through which the organisational networks identified above were formed and in which they interacted with their component organisations. Tools to do this already exist; in particular, network analysis (Knokke and Kuklinski, 1984) may have important insights to offer. To date this technique has been used in a largely descriptive fashion but can surely offer more evaluative insights. This paper should be seen, therefore, as the start, not the conclusion, of our understanding of the relationship between VNPOs, their innovative capacity and the external environment.

References

Abernathy, W., Clark, K. and Kantrow, A. (1983) *Industrial Renaissance*. Basic Books, New York.

Aiken, M. and Hage, J. (1974) "Organic organizations and innovations," *Sociology*, **5**, 63–82.

Alter, C. and Hage, J. (1993) *Organizations Working Together*. Sage, Newbury Park.

Ascher, K. (1987) *Politics of Privatisation*. Macmillan, London.

Astley, G. and Van de Ven, A. (1983) "Central perspectives and debates in organization theory," *Administrative Science Quarterly*, **28**, 245–273.

Beekum, R. and Ginn, G. (1993) "Business strategy and inter-organizational linkages within the acute care hospital industry: An expansion of the miles and snow typology," *Human Relations*, **46**(11), 1291–1312.

Benson, J. (1975) "Inter-organizational networks as a political economy," *Administrative Science Quarterly*, **20**(3), 229–249.

Bessant, J. and Grunt, M. (1985) *Management and Manufacturing Innovation in the United Kingdom and West Germany*. Gower, Aldershot.

Best, M. (1990) *The New Competition*. Polity Press, Cambridge.

Beveridge, W. (1948) *Voluntary Action*. Allen and Unwin, London.

de Bruijn, J. and ten Heuvelhof, E. (1997) "Instruments for network management," *Managing Complex Networks*, W. Kickert, E-H Klijn and J Koppenjan (eds.), Sage, London, 119–136.

Breiger, R. (1976) "Career attributes and network structure: A blockmodel study of a biomedical research network," *American Sociological Review*, **47**, 117–135.

Camagni, R. (1990) "From 'local milieu' to innovation through co-operation networks" *Innovation Networks, Spatial Perspectives*, R Camagni (ed.), Belhaven Press, New Haven.

Collins, H. (1974) "The TEA set: Tacit knowledge and scientific knowledge," *Science Studies*, **4**, 165–185.

Commission on the Future of the Voluntary Sector (1996*) Meeting the Challenge of Change*. NVCO, London.

Conway, S. (1997) "Strategic personal links in successful innovation: Link pins, bridges and liaisons," *Creativity and Innovation Management*, **6**(4), 226–233.

Child, J. (1972) "Organizational structure, environment and performance: The role of strategic choice," *Sociology*, **6**, 1–22.

de Bresson, C. and Amesse, F. (1991) "Networks of innovators: A review and introduction to the issue," *Research Policy*, **20**(5), 363–379.

Department of Health (1989) *Caring for People*. HMSO, London.

Department of the Environment Transport and the Regions [DETR] (1998) *Community-Based Regeneration Initiatives — a Working Paper*. DETR, London.

Deutsch, K. (1985) "On theory and research in innovation," in *Innovation in the Public Sector*, R. Merritt and A. Merritt (eds.), Sage, Beverley Hills, 17–34.

Ferlie, E., Ashburner, L, Fitzgerald, L. and Pettigrew, A. (1997) *The New Public Management in Action*. Oxford University Press, Oxford.

Freeman, C. (1982) *Economics of Industrial Innovation*. Frances Pinter, London.

Gomulka, S. (1990) *The Theory of Technological Change and Economic Growth*. Routledge, London.

Granovetter, M. (1985) "Economic action and social structure: A theory of embeddedness," *American Journal of Sociology*, **91**(3), 481–510.

Griffiths, R. (1988) *Community Care: Agenda for Action*. HMSO, London.

Herbig, P. (1991) "A cusp catastrophe model of the adoption of an industrial innovation," *Journal of Product Management Innovation*, **8**(2), 127–137.

Home Office (1990) *Efficiency Scrutiny of Government Funding of the Voluntary Sector*. HMSO, London.

Hood, C. (1991) "A public management for all seasons," *Public Administration*, **69**, 3–19.

Johnson, N. (1987) *The Welfare State in Transition*. Wheatsheaf, Brighton.

Jones, O. (1997) "Structuration theory and technology transfer: Toward a social theory of innovation," *Aston Business School Research Institute Working Paper*, *RP9704*, Aston University, Birmingham.

Kickert, W. (1993) "Complexity, governance and dynamics," *Modern Governance*, J. Kooiman (ed.), Sage, London, 191–204.

Kickert, W. (1997) "Public governance in the Netherlands: An alternative to Anglo-American managerialism," *Public Administration*, **75**(4), 731–752.

Kickert, W. and Kopperjan, J. (1997) "Public management and network management," *Managing Complex Networks*, W. Kickert, E-H Klijn and J. Koppenjan (eds.), Sage, London, 35–61.

Kickert, W., Klijn, E-H and Koppenjan, J. (1997) "A management perspective on policy networks," *Managing Complex Networks*, W. Kickert, E-H. Klijn and J. Koppenjan (eds.), Sage, London, 1–13.

Kimberly, J. (1981) "Managerial innovation," *Handbook of Organizational Design*, P. Nystrom and W. Starbuck (eds.), Oxford University Press, Oxford, 84–104.

Kings Fund Institute (1987) *Promoting Innovation in Community Care*. Kings Fund Institute, London.

Klijn, E-H (1996) "Analyzing and managing policy processes in complex networks: A theoretical examination of the concept policy network and its problems," *Administration and Society*, **28**(1), 90–119.

Klijn, E-H., Koppenjan, J. and Termeer, K. (1995) "Managing networks in the public sector: A theoretical study of management strategies in policy networks," *Public Administration*, **73**, 437–454.

Klijn, E-H. and Tiesman, G. (1997) "Strategies and games in networks," *Managing Complex Networks*, W. Kickert, E-H Klijn and J. Koppenjan (eds.), Sage, London, 98–118.

Knapp, M., Robertson, E. and Thomason, C. (1990) "Public money, voluntary action: Whose welfare?" in *The Third Sector*, H. Anheier and W. Seibel (eds.), de Gruyter, Berlin, 183–218.

Knokke, D. and Kuklinski, J. (1982) *Network Analysis*. Sage, Beverley Hills.

Kramer, R. (1981) *Voluntary Agencies in the Welfare State*. University of Berkeley Press, California.

Kreiner, K. and Schultz, M. (1993) "Informal collaboration in research and design: The formation of networks across organizations," *Organizational Studies*, **14**(2), 189–209.

Labour Party (1997) *Building the Future Together*. Labour Party, London.

Lane, J-E. (1993) *The Public Sector*. Sage, London.

Lynn, L. (1996) "Assume a network: Reforming mental health services in Illinois," *Journal of Public Administration Research and Theory*, **6**, 297–314.

Mellor, M. (1985) *Role of Voluntary Organisations in Social Welfare*. Croom Helm, London.

Miles, R. and Snow, C. (1978) *Organization Strategy Structure and Process*. McGraw Hill, New York.

Milward, H. B. and Provan, K. (1997) "Principles for controlling agents. The political economy of network structure," *Second International Research Symposium on Public Services Management*, Aston University, Birmingham.

Ministry of Health (1959) *Report of the Working Party on Social Workers in the Local Authority, Health and Welfare Services*, HMSO, London.

Mischra, R. (1984) *The Welfare State in Transition*. Wheatsheaf, Brighton.

Nelson R. (1993) "Technological innovation: The role of non-profit organizations," in *Non Profit Organizations in a Market Economy*, D. Hammack and D. Young (eds.), Jossey Bass, San Francisco, 363–377.

Osborne, D. and Gaebler (1992) *Reinventing Government*. Plenum, New York.

Osborne, S. (1996) "Selecting a methodology for management research: Issues and resolution," *PSMRC Working Paper 34*, Aston University, Birmingham.

Osborne, S. (1998a) "Naming the beast: Defining and classifying service innovations in social policy," *Human Relations*, **51**(9), 1133–1154.

Osborne, S. (1998b) *Voluntary Organizations and Innovation in Public Services*. Routledge, London.

Osborne, S. (1998c) "Partnerships in local economic development: A bridge too far for the voluntary sector?" *Local Economy*, **12**(4), 290–295.

Osborne, S. (1999) *Promoting Local Voluntary and Community Action*. YPS, York.

Osborne, S. and Hems, L. (1995) "The economic structure of the charitable sector in the United Kingdom," *Non Profit and Voluntary Sector Quarterly*, **24**(4), 321–336.

Osborne, S. and Ross, K. (1999) "Managing the policy — practice interface in government — nonprofit relations. The case of area regeneration in the UK," *Right Conditions for the Development of Nonprofit Organizations Congress*, Parma.

Osborne, S. and Tricker, M. (1994) "Local development agencies: Supporting voluntary action," *Non Profit Management and Leadership*, 5(1), 37–52.

O'Toole, L., Hanf, K. and Hupe, P. (1997) "Managing implementation processes in network," *Managing Complex Networks*, W. Kickert and E-H. Klijn and J. Koppenjan (eds.), Sage, London, 137–151.

Peyton, R. (1989) "Philanthropic values," *Philanthropic Giving*, R. Magat (ed.), Oxford University Press, New York.

Pfeffer, J. and Salancik, A. (1978) *The External Control of Organizations*. Harper Row, New York.

Porter, M. (1985) *Competitive Advantage*. Free Press, New York.

Poulton, G. (1988) *Managing Voluntary Organisations*. John Wiley, Chichester.

Rao, N. (1991) *From Providing to Enabling*. Joseph Rowntree Foundation, York.

Rhodes, R. (1990) "Policy networks: A British perspective," *Journal of Theoretical Politics*, 2(3), 293–217.

Rhodes, R. (1997) *Understanding Governance*. Open University Press, Buckingham.

Rogers, E. (1987) "Progress, problems and prospects for network research: Investigating relationships in the age of electronic communication," *VII Sunbelt Social Networks Conference*, Florida.

Salamon, L. (1995) *Partners in Public Service*. Johns Hopkins UP, Boston.

Singh, J., Tucker, D. and Meinhard, A. (1991) "Institutional change and ecological dynamics," *The New Institutionalism in Organizational Analysis*, W. Powell and P. DiMaggio (eds.), University of Chicago Press, Chicago, 390–422.

Smale, G. and Tuson, G. (1990) "Community social work: Foundation for the 1990s and Beyond," in *Partners in Empowerment: Networks of Innovation in Social Work*, G. Darville and G. Smale (eds.), NISW, London.

Tucker, D., J. Baum and Singh, J. (1992) "The institutional ecology of human service organizations," in *Human Services as Complex Organizations*, Y. Hasenfeld (ed.), Sage, Newbury Park, 47–72.

Tushman, M. and Anderson, P. (1985) "Technological discontinuities and organizational environments," *Administrative Science Quarterly*, **31**, 439–465.

Urabe, K. (1988) "Innovation and the Japanese management style," in *Innovation and Management: International Comparisons*, K. Urabe, J. Child and T. Kagono (eds.), de Gruyter, Berlin, 3–76.

Wagner, G. (1988) *Residential Care, A Positive Choice*. HMSO, London.

Webb, S. and Webb B. (1911) *Prevention of Destitution*. Longman, London.

Wistow, G., Knapp, M., Hardy, B. and Allen, C. (1994) *Social Care in a Mixed Economy*. Open University Press, Buckingham.

Wistow, G., Knapp, M., Hardy, B., Forder, J., Manning, R. and Kendall, J. (1996) *Social Care Markets: Progress and Prospects*. Open University Press, Buckingham.

Wolfenden Committee (1978) *Future of Voluntary Organisations*. Croom Helm, London.

Zahria, S. and Pearce, J. (1990) "Research evidence on the Miles-Snow typology," *Journal of Management*, **16**(4), 751–768.

Chapter 10

Innovation Through Postmodern Networks: The Case of Ecoprotestors

David Crowther and Stuart Cooper

Introduction

Arguably, one of the most significant developments in the study of anthropology was brought about through the technique of mapping interactions and relationships between the people, traditions and cultures of primitive tribes (Marcus and Fischer, 1986). Such techniques were employed from the early part of the twentieth century to study and map the interactions of primitive tribes in various part of the world, such as the Trobiand Islands (Malinowski, 1922), Samoa (Mead, 1949) and the Amazon basin (Levi Strauss, 1961). In this paper, the authors argue that the understanding of organisational activity can still be enhanced through the study of different cultures. However, rather than seek out so-called primitive cultures the authors base their study upon an alternative culture present in British society, and its interaction with more conventionally constituted society. From this, we argue that such a study can enhance our understanding of network theory and its role in innovation. Thus, this study is concerned with the alternative culture of the New Age Travellers (referred to subsequently as 'travellers' — the term they themselves use) and their interaction with local communities via their involvement in the ecoprotest movement.

The Ecoprotest Movement

The ecoprotestor movement has been in existence in the UK for a number of years. The phenomenon has been manifest as a direct action protest against the destruction of nature for the construction, in most cases, of roads, although one particularly high profile protest was concerned with the extension of Manchester airport. However, not all protests of this nature are given such high profile and many remain merely local phenomena creating effects and news only within a small spatial environment.

Despite their profile as direct action protestors, ecoprotestors have received a mostly favourable press. Their impact upon the geographical landscape of the country has however been largely ephemeral and has been restricted to the abandonment of some small-scale projects and the delaying of proposed developments. Indeed, in the main, they define their success for this form of protest in terms of the length of time by which they delay such developments, the increased cost of these developments (brought about by the increased security and eviction costs), and the extent to which the environmental issues involved have been publicly aired (Doherty, 1997). In this respect, they realistically adopt a long term perspective on their form of protest and take the view that each incident, although ultimately ending in defeat, is merely one step upon the road to eventual victory in giving long term environmental impact precedence over short term economic gain. These views are openly expressed by many of the ecoprotestors involved in these protests and therefore, in this respect, it can be claimed that these seemingly transient individuals have a strategy and organisational objectives (Johnson and Scholes, 1997) which are as coherent as any formally constituted organisation. What these people do not have, however, is a formal organisational structure and procedures which equates to those of more conventionally constituted organisations. Consequently, the way in which these protesters construct their objectives and operationalise them into activity forms an interesting study from which more formally constituted organisation can potentially learn.

The study evaluates the activities of the ecoprotestors from a network perspective. This approach was chosen because the study identifies multiple communication levels and network analysis provides 'a potential framework' for such research (Euske and Roberts, 1987; Steward and Conway, 1998). Specifically, two different types of network are identified. The first of these is the traveller network; this study is primarily concerned with the highly effective news dissemination network that exists amongst the geographically re-mote sites. According to the ecoprotestors interviewed, there are approximately 12–20 such protest sites in existence at any particular point in time; the majority of these are known only to the ecoprotestors themselves and to the inhabitants of the local area of any particular protest site. The second network analysed is that which links these travellers to members of conventional society, in other words, those individuals in a given local community affected by a proposed development that has been targeted by an ecoprotest action. These different networks of communication may be classified as either internal to the community (an inter-group or intra-organisational network), or external to the community (an inter-organisational network), in the terminology of Fulk and Boyd (1991).

This research is based upon interviews with, and non-participant observation of, two particular groups of protestors:

- those involved in the protest against the Birmingham North Relief Road (BNRR) and based in the first camp to be established in woods alongside the A38 near Tamworth;

- those involved in the protest against road developments infringing upon Bass's Recreation Ground, a public park very close to the centre of Derby.

It is argued in this paper that the ecoprotest movement is an exemplar of a truly postmodern network. In order to explore this argument however, in terms of its relevance to the study of organisations, it is first necessary to consider the nature of the postmodern environment for such organisations.

The Postmodern Environment of Organisations

In considering representations of the performance of organisations in a postmodern environment it is necessary first to discuss the nature of postmodernity. It has been considered from a variety of perspectives: for example, as being epochal (Collins, 1989), in replacing modernity as the current time frame; or epistemological (Newton, 1996), in its relativity to other interpretations of social structures, or a negation of modernity (Featherstone, 1988). The concept of postmodernity is however considered most usefully by Lyotard (1984), who questions the use of modernist meta-narratives which legitimate society as existing for the good of its members with the consequent presumption that the whole unites the parts as an expression of the common good; it is this latter definition of postmodernity that is adopted in this paper. Thus the meta-narrative of economic rationality legitimates both the existence of organisations and the classical liberal approach which assumes that the free market provides a mediating mechanism which ensures that the freedom of organisations to pursue their own ends will inevitably become synonymous with that freedom leading to optimal benefit for both the owners of that organisation and for the other stakeholders to that organisation.

The postmodern environment has been argued (Crowther and Duty, 1996) to be one in which organisations are becoming increasingly susceptible to temporal pressures, brought about by the compression of space and time. Thus, business organisations have found themselves having to adjust many of their operational features in order to cope or just to survive. The crisis that this 'new age' has brought about seems to have become an accepted part of life. This is manifest particularly in the fact that business organisations accept that they have to adapt, to become more organic and flexible. It is generally accepted that the speed of change will continue to increase in the future, and that what will be required of organisations is the ability to create flexible structures in order to operate effectively. Harvey (1990) argues that the compression of space and time has been brought about through development in the technological and informational architecture of society. This has the effect of removing the imperative for territorial boundaries from organisations and thus providing an opportunity for the redefinition of the concept of organisation in terms of organising local structures for the provision of local goods and services. The implication of this is that organisational structures need no longer be dictated solely by the need for transaction cost minimising models of service provision and the ability to define themselves afresh for the provision of individual goods and services becomes possible. The traveller network, as a truly postmodern form of organisation, provides an example of this in practice, as will be seen in the subsequent analysis.

This redefinition of organisation contains within itself one of the inherent contradictions of a postmodernist view of the world, namely that between the borderlessness of any organisation and the extreme territorial inclusion/exclusion criterion adopted for performance evaluation and reporting systems within the boundary or the organisation as a whole. Therefore the success of organisations, as measured by performance indicators, usually of the accounting variety (already charged with irrelevance due to their limitations (Eccles, 1991)) ignores several crucial factors of that performance in a postmodern environment (Crowther and Duty, 1996). This has the effect of polarising organisations and distancing them from a unified

focus in their operating and reporting structures as the organisational boundary collapses in significance, and to expand the concept
to inclusion in an expanded environment for some purposes while
at the same time shrinking the concept of locality of operations for
other purposes (Radhakrishnan, 1994). This interpretation suggests
that different spaces are needed for different histories and that a
dominant model of the organisation within society has no rational
meaning. When considering the question of organisations, the identity
of constituents of such an organisation, and their relationship with
the macro culture and with societal structure, this implies that the
local structure has dominant importance to the individual constituents of the organisation and that any sense of community, in terms
of operating environment, is defined circumstantially. Thus, an individual considers him/herself to be a member of an organisation as
a community for a particular purpose and a member of a different
organisational community for different purpose, with this identity
being defined in terms of commonality of interest for specific purposes rather than being an overriding part of any definition of the
organisation. It is in this way that the travellers and ecoprotestors
are able to opt into and out of each community as they wish, while
the community as an organisation maintains its continuous temporal
existence.

A postmodernist view of organisations and the nature of
their behaviour is that they are sustained by the rules governing
their existence and by the resource appropriation mechanisms
which apply to them, rather than by any real need from the people
who they purport to serve (Barnett and Crowther, 1998). Thus the
legitimation of their very existence is not founded upon this redefinition of organisational identity and community need. Rather this
redefinition of community, for organisational and transaction enaction
purposes, suggests that a very different type of organisational structure
is needed, and indeed exists, in order to cater for the needs of the
individuals constituents of that organisation who aggregate for one
common purpose while atomising (or aggregating with different

individuals) for others. Such a structure of organisations has been defined by Heckscher (1994) as 'post-bureaucratic', with its rationale for continuing existence not being through self-referential normalising mechanisms but rather through the maintenance of an interactive dialogue, based upon consensus, with the individual members of the stakeholder community which the organisation exists to serve. Thus, the traveller network acts as a dialogue machine both within the network itself and through its interface with other stakeholders when acting in ecoprotest mode. This organisational structure can be extended to exclude a territorial basis for existence (Nohria and Berkley, 1994) whereby the organisation, through the use of informational and communicational technology, need be little more than a virtual organisation existing in a virtual environment as the need arises. Thus, the continuing existence, either temporally or geographically, of any organisation as a unit of service provision has no meaning in its own right, as the organisation has no purpose other than the provision of the functions mandated to it by the stakeholder community, in its widest definition, which it serves. Such an instrumental view of organisations and their constituent parts would be radically different from existing paradigms and inter-pretations but this would be fully consistent with any postmodernist definition based within the concept of the revised stakeholder community.

Similarly, Popper (1945) argues that present trends do not necessarily continue into the future and that any amount of empirical evidence and economic or sociological analysis does not change the lack of predictive power of past data. For a self-organising society of course this is not an issue as actions arise without any need for forward planning and the prediction of the future. Therefore, in this respect, it is argued that the postmodern networks under consideration are inherently more stable than traditional organisations.

The argument of Derrida (1978) provides the motivation and means to integrate the analysis of strategy formulation, and the operational activity necessary, to the successful operating of an organisation in this postmodern environment with the requirements of the organisation as a whole, as manifest for the need for traditional decision making procedures. Thus, rather than seeking to develop two independent operational structures — postmodern at an operational level and traditional at an organisational level — it is desirable to integrate these two structures into one to meet all needs. This can be achieved through a recognition of the working of a self-organising postmodern network to become manifest in the organisational structure and procedures, thereby recognising that the continued existence of the organisational boundary, deemed irrelevant to any postmodernist analysis, is a crucial feature of any modernist inter-pretation of the organisational environment. This can be illustrated through our consideration of the ecoprotest phenomenon.

The Ecoprotest

A common feature of ecoprotests is the need for a permanent site to be established that acts as a focal point for the protest. It is perhaps significant that in the cases considered in this paper, and in the vast majority of, if not all, other cases, the protest site has been established by a member of the local community. This person, in order to establish such a protest site, must inevitably be familiar with this form of protest and be in touch with others who are engaged in such protest. In fact, this initiator can be classified as being a member of both networks explored within this paper. The first network is that of the traveller who is involved full time in either protest or in travelling. Contact, from the initiator, with such travellers is essential as a permanent site requires people to be permanently resident and it is only travellers who can fulfil this role in the protest. The second network is that which exists between the travellers and the people external to it, but most importantly that of the local community affected by the proposed development. These members of the local community can be described as part-time protestors as they will not be permanently resident at the protest site. The distinction between the full-time ecoprotestor/traveller and the part-time ecoprotestor/local resident is an important one and gives rise to the two levels of network analysis undertaken in this research. For ease of reference, these two networks will be referred to as the traveller network and the local community network, respectively, in the remainder of this study.

At any individual site the number of people who are present at any particular time vary, and will comprise a number of permanent residents, travellers, and a number of part time protestors, local residents. Thus, it was possible for the researchers to visit a camp at one time and find only two people present while at another time there were approximately 40 people present. Local inhabitants will tend to be present outside working hours and to be staying at the camp only during weekends and holidays. Travellers themselves are in the main 'static', that is, it is common to find that they are permanently resident at one camp for a period of time before moving on to either another protest site or another traveller site. Very few remain at a site throughout the duration of the protest although there seem always to be a (very small) number of dedicated protestors who will remain throughout the period of the protest. This is not however arranged in any formal manner and it would seem to be reasonable, given the transience of all protestors, to find that at a particular point in time there is no one in occupation at a particular protest site. The fact that this never happens indicates the existence of a form of organisation that transcends traditional structures, a point that will be returned to later.

The protestors themselves actively encourage visits from anyone and the fostering of relationships with local inhabitants and anyone else affected by the proposed development is viewed as an important part of their role and an important determinant of their chance of a successful protest. Thus, visitors are actively encouraged and hospitably received, and in the case of the BNRR site steps were taken to publicise the protest to road users who might be considered to be beneficiaries of the proposed development. Thus, the protestors are open and fully prepared to talk about their activities, their experiences, and their objectives, to any interested person. Significantly, they are also familiar with academic research and used to being the subject of such research; indeed when the authors first visited one camp and during initial conversations mentioned that they were academics the question was immediately asked as to what we were researching! However, being the subject of research had no discernible impact upon their preparedness to answer questions and talk about themselves.

As noted earlier, there are two distinct networks that will be considered in this chapter — the traveller network and the protest network. Each of the two networks are involved in a specific protest to a certain extent, either through a presence on the protest site or through other means. The protest camp itself can be likened to a nexus where these two networks interact. The travellers themselves constitute as an interacting community spatially separated over several different camps. Although spatially separated the travellers themselves, as individuals, both move between camps and have an informal communication network. This is illustrated in Fig. 1 below, where all

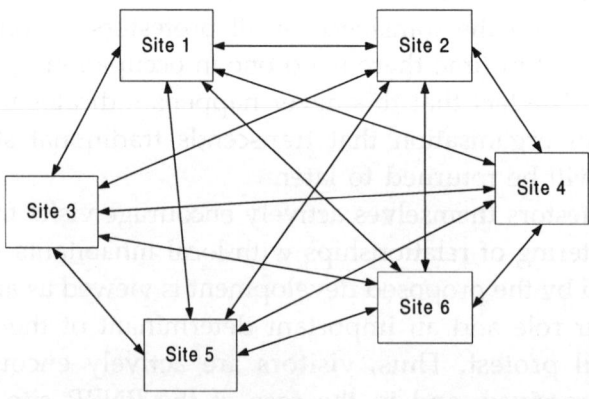

Fig. 1. The traveller community network.

relationships are depicted as positive but of varying strengths, depending upon the number of people at each site and the degree of interaction.

Moreover, travellers become ecoprotestors when they choose to become involved in such protests and in this mode they also belong to a different network — the ecoprotest network, in which they are joined by members of the local community. It is this network which has an interaction with the developers of the site in question, and it can be depicted thus as in Fig. 2:

These figures incorporate the key relations within the two networks. In the terms of Scott (1991), as derived from the work of Cartwright and Harary (1956), some of the relations depicted are undirected, as it is considered that the relationship between the different nodes are reciprocal, and are signed as dependent upon the positive or negative relation between the nodal points. Because these networks are 'postmodern' (see later) it is also important to note that the strength of the ties cannot be depicted as they fluctuate in strength both temporally and spatially according to the desires of the members of the network, and even at times will disappear.

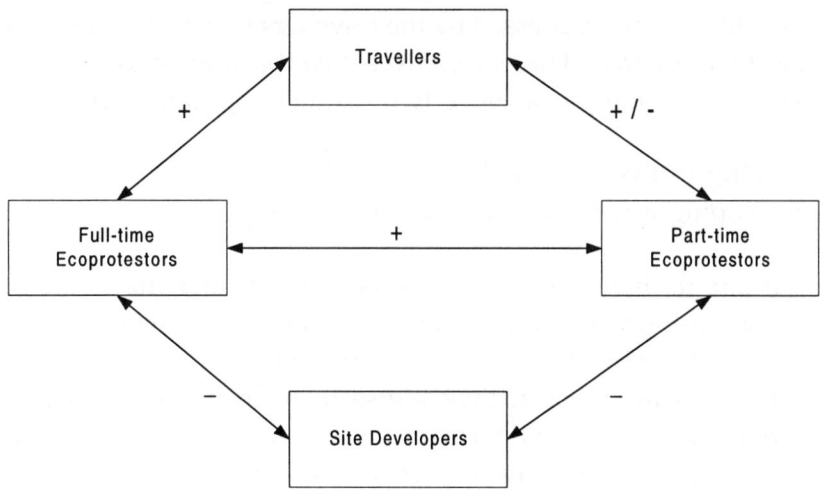

Fig. 2. The protest network.

In order to develop an analysis of the significance of the ecoprotestor phenomenon to organisational behaviour it is necessary to first develop an understanding of the strategic goals of the travellers, of which the full-time ecoprotestors are a subset, and then to consider for each of the two networks the extent of their actions and interactions. From this analysis, a fuller understanding of the protest and its effects can be derived, and from the workings of these network analogies lessons will be drawn for use within all modern organisations.

The Strategic Goals of New Age Travellers

The traveller community has specific strategic objectives. These objectives are not explicitly stated and have not been arrived at by any overt decision making; rather they have developed over time through an unconscious process. It is argued here that this unconscious process is one of the adaptive features of the kind of self organising system which the travellers represent and this is a point considered in more detail later. Nevertheless, these strategic objectives are clearly defined and openly expressed by the travellers who were interviewed during this research. These objectives have been expressed by these travellers as seeking to achieve two distinct objectives, which are:

- learning to live differently
- developing a community spirit and identity

By learning to live differently these people are referring to the need for a relationship with nature which must not be driven by the normal societal motives of economic consumption and wealth creation. Therefore, in this respect, they refuse to recognise the proprietary ownership mores of mainstream society as a basis for resource utilisation, and this applies in particular to land which is viewed as a common resource to be used rather than owned. Thus, their view of the economic drive towards ownership of land can be considered to be in common with both North American Indians (Driver, 1961; Fagan, 1995; Gill, 1982) and the seventeenth century Digger movement in England (Aylmer, 1975; Pease, 1965). This view of land and its use is of course one of the principal reasons why the traveller movement has so often been in conflict with conventional society and why mainstream publicity about them has been uniformly bad. As with the Digger movement, their very existence, and possible survival, can be considered a threat to the economic basis of societal existence. The travellers themselves would be pleased to be perceived in this way as this provides a level of support for their ideas of how individuals should live out their life. It is perhaps worth noting, however, that the Diggers were violently suppressed in the seventeenth century and North American Indian views were constantly ignored in the exploitation of North America (Brown, 1991; Josephy, 1995). The future of the traveller movement and its philosophy might therefore be considered to be questionable. Equally, other nomadic groups have been subject to persecution for their way of life in the same manner as travellers (Fonseca, 1996).

Not only have these views about the economic imperative been adopted by travellers, but also some of the technologies of North American Indians; thus, at any traveller site tee-pees can be seen. Equally, the other main form of housing adopted by travellers is 'the bender', a technology which has been acquired from other nomadic groups. Communal living and the sharing of food and resources are also important features of the traveller way of life that can be found among such groups. For the travellers, these are all essential parts of their way of life which are viewed as forming part of their strategic objective of developing community spirit and alternative ways of living, all of which constitute a rejection of current societal modes of organisation. These views can be found expressed to a greater or lesser extent amongst travellers, who, in the experiences of the authors, are generally thoughtful and well-educated people rather than refugees from society, as common mainstream views would depict them. It is important to note that although ecoprotestors have received a mostly favourable press the travellers themselves are less well-received.

The other main strand of traveller philosophy, which can be seen from their way of life and their general involvement in the ecoprotest movement, is a general concern with the environment and its degradation through developments. This is particularly true when the proposed developments are for the purpose of increasing road or air transport at the expense of nature. In this they are of course not alone and this seems to be the principal reason why the ecoprotest movement enjoys popular support and involves a range of people with no concern other than environmental protection. It also perhaps explains the existence of other diverse groups such as the mainstream Transport 2000 group and the fringe group, Reclaim the Streets. However, such ecoprotests are not restricted to transport related developments, but increasingly are also concerned with protesting against genetically modified crops and the ensuing environmental damage (Garret, 1994). In such protests, it is argued that travellers, and other ecoprotestors, are acting both in the capacity of the societal conscience, through the taking of actions which others agree with but will not participate in, and the focus for consensual democracy, through publicising the dangers of such developments. Indeed it can be argued that the government accepts this rationale for ecoprotest and is in the process of developing alternative, more environmentally friendly and sustainable, means of economic development.

The Traveller Network: A Self-Organising System

As noted earlier, the only people who can realistically be permanently resident at an ecoprotest site are travellers. The new age travellers in the UK can be considered to exist both as a community and as a network which is continually constituted and reconstituted by the way in which the network acts as a dialogue machine, fostering communication within the network between members of the community and others outside the community. As a community the traveller network must be viewed as a postmodern community (Crowther and Carter, 1998; Barnett and Crowther, 1998) which has neither a contiguous spatial existence nor a contiguous temporal existence, but nevertheless has both a spatial and temporal presence as needed and desired by the community. This type of community has no fixed membership but comprises those people who have chosen to be members at any particular point in time. This is distinct from the local neighbourhood communities in which a network of ties is inherent (Tilley, 1974; Wellman and Leighton, 1979). It is these factors of spatial and temporal discontinuity and self-determination of membership which mark these communities and their networks as truly postmodern, particularly when coupled with their ability to exist, organise and develop without any organisational structure. What is significant from the point of view of organisations is that these networks are able to achieve action without any need for organising and in this respect they do not merely mimic nature through their ability to survive and adapt but also parallel human organisations but with the formal mechanisms for action removed — actions without transactional costs!

The operating of the traveller community can be considered as a network of weak ties (Granovetter, 1973) which are based upon social interaction around a commonality of interests rather than based upon friendship. In this respect, it is pertinent to note that the travellers interviewed appear to be well-educated people, which supports the findings of Ericksen and Yansey (1980) that strong ties are made more use of by less well-educated people. Equally, Pool (1980) demonstrates that networks operate more effectively when an effective communication system exists. The postmodern nature of this network, manifest through its lack of spatial and temporal contiguity, can be argued to increase the effectiveness of its achievement of the strategic goals expressed earlier. However, it can also be seen to actually increase the effectiveness of the network as a communication mechanism and it is to these that we now turn.

The mechanisms mobilised to promote communication enables the traveller network to act as a dialogue machine. Although they have their own press, which is operated by members of the community, the prime means of communication between travellers, and between travellers and protestors, is word of mouth. This relies upon the essentially postmodern nature of the community network whereby spatial location is not fixed but rather is in a continual state of flux as the community members enter, leave and move between sites. This is represented in Fig. 1, by the fact that each site is connected to all of the other sites. This is operated on a completely informal basis, without any organisation, but nevertheless operates sufficiently effectively for knowledge of the various traveller sites and the various protest sites to become common knowledge to travellers without the need for written communication. In addition to this flow of knowledge via the movement of individuals between sites, the various spatially disparate sections of the community will 'keep in touch' through meeting at certain common interest events, such as festivals and demonstrations; due to this common interest and the knowledge that other members of the community will be in attendance, these events are well-attended. Other prime communication events are centred around the various festivals which the community, or rather individuals within the community, organise. These too are arranged without overt publicity but become common knowledge within the community, and hence are well-attended, through the effectiveness of the informal communication network of which this postmodern community comprises. Therefore, in this respect, it can be argued that the traveller communication network operates at a greater level of effectiveness, if not efficiency, than do traditional organisational communication mechanisms. In this sense, effectiveness is defined as a message being received and understood as intended by the recipients (Simons and Naylor-Stables, 1997).

The communication among the protestors and travellers takes place in an entirely uncoordinated manner and as such the network that facilitates this communication can be considered to be a self-organising system. Self-organisation has been described as 'a process in which the components of a system in effect spontaneously communicate with each other and abruptly co-operate in a co-ordinated and concerted common behaviour' (Stacey, 1993:240). As such the network can be considered to have the characteristics of a complex adaptive system (Vriend, 1994; Phelan, 1995); a complex system is one in which many agents interact with each other and it is adaptive if the agents' actions alter as a result. In fact, this is an exemplar of a self-organising system that protects the welfare of vulnerable groups and therefore refutes the argument of Bovaird and Sharifi (1998) concerning the potential dangers of such systems to these groups.

This argument concerning the effectiveness of travellers and protestors in operating as self-organising systems can be further demonstrated through a consideration of their mode of organising and operating to achieve their strategic goals. In this respect, the protest sites have no formal organisation and no plan of operational activities but nevertheless things get done as they need to be done. This requires individuals within each site to assume responsibility for undertaking tasks, and co-operating as required. In the experience of the protestors (supported by the observations of the researchers) this happens effectively. Ecoprotestors are well-known for their ability to dig tunnels and build tree houses safely and this too derives from individuals within the protest who have the necessary expertise and are willing to teach others. For example, on one site which we visited there was a blacksmith who was able to make doors for the tunnels, while on the other site another individual had the necessary expertise to determine that tunnel building was not practical. Some of these skills, such as smithying, can easily be acquired from elsewhere and transported into ecoprotest. However, this is not the case for tree house and tunnel building, and such skills are most likely to have originated from within the ecoprotest movement. McAdam (1983) has argued that protest techniques are diffused through time from one movement to another, whereas Tarrow (1995) has argued that new forms of protest tend to arise in the early stages of a protest movement and then become part of the general repertoire of such protests. In no case however can a mechanism for the invention of such tactics be identified and this evolutionary nature of the tactics of protest can be seen to be one of the distinguishing features of this form of self-organisation that can be contrasted with the formal R&D procedures of traditional organisations. For most travellers, however, knowledge of how these skills developed is unknown and has simply been accepted into the repertoire of ecoprotest. Indeed we were unable to speak with anyone who had experience of the development of these skills, thereby providing an example of spontaneous innovation and adaptation by this self-organising system.

The Local Community Network

All ecoprotest is pacifist, but it needs to be recognised that the protest involves a degree of illegitimacy in its means. Indeed, the discourse surrounding environmental protest is one of illegitimacy, depending upon whether one considers that the ends justify the means or not, and their impact upon legitimate organisations, including government, tends initially to be one of increasing transaction costs for the firms targeted with a corresponding long term change in decisions made and actions taken. However, Chaliand (1987) has argued that a successful protest organisation, grounded in illegitimate activity, needs a base in society which extends beyond its membership, and requires popular support in order to exist and achieve results. The ecoprotestors are well-aware of this need for support from society and regard it as necessarily being manifest in the local community. Thus, they regard good relations with the local community as extremely important, both with regard to their continuing existence and for the success of their protest. An important point to consider, as depicted in Fig. 2, is the imbalance of the network. The societal perception of travellers, from which full time ecoprotestors are taken, is stereotypically negative. According to Scott (1991), this imbalance will result in either attempts to change perceptions or place a strain on the network. In this instance, the ecoprotestors are very much using this protest network as a way of attempting to communicate their own, and corresponding traveller, strategic goals and objectives.

The relations with the local community are also important as far as the subsistence of the ecoprotest is concerned. The means to exist at these protest sites seems problematical as full time protestors are unable to draw upon state benefits, as they are patently not seeking work. Indeed, such protestors do not seek to exist upon state benefits and instead rely upon donations from supporters, within the local community, as a means of existence. Such donations take the form of money, materials and food, and each site freely publicises what they are in need of without exerting any pressure upon visitors to make any form of contribution; rather they are willing to share what they have not just with each other but also with any passing visitor, such as ourselves, who drops in. Indeed money, food and site materials seem to be held communally. The protestors however seem to have no worries concerning their ability to feed and support themselves, and describe the support provided by supporters in the local community as both welcome and generous. Such support extends to such things as local inhabitants inviting them into their houses for a bath. The support provided by the local community inevitably depends to some extent upon the geographical location of the site. This is not because the reception from the local community is less welcoming in some areas — indeed they describe the support everywhere as positive — but rather the proximity of the camp to local inhabitants. In this respect, the site in Derby, located close to the city centre, differs from the BNRR site, which is situated some miles from any local inhabitants. Thus, the protestors at the Derby site have benefited from such things as benefit concerts held in local pubs as well as favourable publicity in the local newspaper, the Derby Evening Telegraph. In contrast, the BNRR protestors have had no such support and have had to rely upon locals making an active decision to visit the site and provide donations. Nevertheless, even this site has had no problem in receiving such support and necessary donations. From one perspective, such a protest site may seem parasitic as it cannot be sustained without an outside injection of resources; from another perspective, however, this demonstrates the existence of a true community with the mutual interchange of goods for services without any need for formal mechanisms or any ongoing commitment.

The local community provides essential resources to the ecoprotestors, but it is the existence of an ecoprotest site that acts as a catalyst for the formation of local protest in a variety of ways. The site seems to spark existing local groups, such as civic societies, into formalising protest and taking a variety of actions; it appears to act as a catalyst for the formation of local groups to undertake protest; and it also seems to act as a mechanism for the mobilisation and focusing of public pressure against the proposed development. This is manifest not just in the support given to the ecoprotestors, but also in other actions taken by the local community. The ecoprotestors, being fully aware of the need for local support and local action, tend to participate in these local actions, although not all ecoprotestors are comfortable with such interaction with the local community and this kind of activity tends not to be uniform but rather left to those who wish to participate. In this respect, community relations activity is similar to all ecoprotestor, and indeed traveller, activity undertaken voluntarily by those who wish to do so. Thus, travellers participate in local protest meetings and groups and in the activities of such groups. Furthermore, the travellers in Derby participated in local activity to a much greater extent and more directly through the organisation of a petition of protest and the collection of signatures throughout the city centre; they also raised the profile of the protest, assisted by newspaper publicity and by the organisation of a public picnic to take place on the protest site and potentially affected area of parkland. In this way, they both mobilised local support and fostered participation in the protest by people of all age groups, thereby nullifying one of the claims of antagonists that such protest is restricted to 'disaffected young people'. In Derby, in particular, this protest was assisted greatly by the support of the local newspaper which undertook its own actions to foster the protest, while at the same time providing media coverage on an almost daily basis, and presumably increasing sales at the same time. However, other local activity, such as the organisation of legal challenges to the proposed development, was initiated by members of the local community without any ecoprotestor involvement.

Innovation Through Networking

It is generally accepted in theories of organisation that the alternative method to decision making based upon hierarchy is one based upon consensus (Wuthnow, 1989). Thus, the alternative to decisions being made at one level in an organisation and being transferred to different levels through authority is based upon all members of the organisation, or a subset of the organisation, collaborating in a decision and then adopting it. However, as the travellers and ecoprotestors act as a self-organising system, and as a postmodern organisation, there can be seen to be an alternative mode of decision making which is independent from the structure of the organisation and dominant modes of operationalising decisions. This third method of decision making and setting and implementing objectives is based upon the complete lack of any formal structure and a complete lack of any process for decision making. Decisions in this type of organisation 'just happen', but the evidence from this research would suggest that this method of decision making is extremely powerful and effective. Of equal effectiveness is the way in which these decisions are implemented. It is argued that this method of decision making is based upon the virtual nature of the organisation, which exists merely as a loose network of people who each have complete freedom to opt in or out of the network and of any decision which has been adopted by the organisation. This points a possible way forward in the structuring of organisation and in the decision making process within organisation, based upon the nature of postmodern organisations.

It has been argued (Dermer, 1988) that organisations consist of a sustained set of beliefs and behaviours and that the existence of organisational rules, beliefs and rituals limit the extent of the control which it is possible for managers to exert in the organisation. Similarly Abernethy and Stoelwinder (1995) have found that formal administrative controls in organisations have needed to be replaced by less obtrusive forms of control due to the increasingly complex nature of the tasks which managers perform. Thus, the control of the decision making domain, as exercised by the dominant coalition of the management team, is constrained by the institutional nature of the organisation and the need to transfer ownership of decisions from the decision makers to the organisation as a whole for the implementation of those decisions. In this context, Covalenski and Dirsmith (1986) state that organisational politics play a key role in the construction of reality as far as members of the organisation are concerned. Thus, although the decision agenda is set by the decision makers, this agenda is in reality constrained by the nature of organisational behaviour and therefore the managers actually have a limited choice of decisions open to them. These decisions are limited by: the available information and the way in which it is presented and interpreted; the organisational rules and rituals which determine the way decisions are put into effect; and the need to transfer decision ownership into the public domain within the organisation. However, for a postmodern virtual organisation such problems do not exist as decisions simply evolve and become adopted and implemented.

Conclusions and Lessons

It has been argued here that there are some significant lessons that can be learnt from the eco-network. It can be seen that the innovative nature of postmodern networks, as manifest in the traveller and eco-protestor networks, can lead to innovations in the general repertoire of protest activities. These arrive through trial and error but then enter the general repertoire of activities through the communication mechanisms. The specific characteristics of the eco-network are the informality of the links, the lack of a hierarchy, the positive relations within the 'organisation' and the numerous and somewhat random nature of the dissemination. Further, in this paper, these communication mechanisms have been considered to be self-organising and existing within a postmodern network. This postmodern network operates so effectively that innovative activities quickly become general activities which are continually refined and adapted through the self-learning ability of such networks. Thus, it is argued that innovative forms of networking lead to innovations in practice. This too is manifest in the decision making and strategy determination mechanisms of such postmodern organisations. These examples provide lessons for more conventional organisations which could be adopted to lead to more effective adaptations to the environment in which such organisations find themselves. In this respect, this is akin to organisational learning (Nonaka and Takeuchi, 1995; Argyris and Schon, 1996) when separated from the power relationships (Coopey, 1994; 1995) inherent in all modern organisations.

Furthermore, it has been argued that such postmodern networks are able to react to an unstable environment more quickly than conventional organisations, and to adapt in an evolutionary manner to the circumstances at the time. They are also willing and able to collaborate with other networks as far as a commonality of self-interest is concerned. It is this ability to adapt when coupled with a decision making process which is separated from organisational structure and from the future prediction imperative of traditional organisations which gives such a network its competitive edge over hierarchical organisations. In this way, it is argued, that such fluid organisations are actually more stable than conventional organisations, which can consequently learn from such self-organising modes of existence.

These lessons from the ecoprotest movement can therefore give a range of suggestions to conventional organisations as to how they can adapt to ensure the continued existence into the future, given the uncertain nature of their environment.

References

Abernethy, M. A. and Stoelwinder, J. U. (1995) "The role of professional control in the management of complex organisations," *Accounting, Organizations & Society*, **20**(1), 1–17.

Argyris, C. and Schon, D. A. (1996) *Organizational Learning II*. Addison Wesley, Wokingham.

Aylmer, G. E. (ed.) (1975) *The Levellers in the English Revolution*. Cornell University Press, New York.

Barnett, N. J. and Crowther, D. E. A. (1998) "Community identity in the 21st century: A postmodern evaluation of local government structure," *International Journal of Public Sector Management*, **11**(6/7), 425–439.

Bovaird, T. and Sharifi, S. (1998) "Partnerships and networks as self-organizing systems: An antidote to principal–agent theory," in *Inter- and Intra- Government Arrangements for Productivity: An Agency Approach*, A. Halmachi and P. B. Boorsma (eds.), Kluwer, London.

Brown, D. (1991) *Bury My Heart at Wounded Knee*. Vintage, London.

Cartwright, D. and Harary, F. (1956) "Structural balance: A generalisation of Heider's theory," *Psychological Review*, **63**, 277–293.

Chaliand, G. (1987) *Terrorism — From Popular Struggle to Media Spectacle*. Saqi, London.

Collins, M. (1989) *Post-Modern Design*. Academy Editions, London.

Coopey, J. (1994). "Power politics and ideology," in *Towards the Learning Company*, J. Burgoyne, M. J. Pedlar and T. Boydell (eds.), McGraw-Hill, Maidenhead, 42–51.

Coopey, J. (1995) "The learning organization, power, politics and ideology," *Management Learning*, **26**, 193–213.

Covalenski, M. A. and Dirsmith, M. W. (1986) "The budgetary process of power and politics," *Accounting, Organizations & Society*, **11**(3), 193–214.

Crowther, D. and Carter, C. (1998) "Community identity in cyberspace: A study of the Cornish community," Research Paper Series, No. RP9826, Aston Business School Research Institute, Birmingham.

Crowther, D. and Duty, D. (1996) "Operational performance in post modern organisations — towards a framework for linking operation measures with traditional measures in the evaluation of performance," paper presented at the *Management Accounting Research Group (MARG) Conference*, Aston Business School.

Dermer, J. (1988) "Control and organisational order," *Accounting, Organizations & Society*, **13**(1), 25–36.

Derrida, J. (1978) *Writing and Difference*. (Translated A. Bass), Routledge & Kegan Paul, London.

Doherty, B. (1997) "Direct action against roadbuilding: Some implications for the concept of protest repertoires," in *Contemporary Political Studies*, Vol. 1, J. Stanyer and G. Stoker (eds.), 147–155.

Driver, H. E. (1961) *Indians of North America*. University of Chicago Press, Chicago.

Eccles, R. G. (1991) "The Performance Evaluation Manifesto," *Harvard Business Review*, **69**(1), 131–137.

Erricksen, E. and Yansey, W. (1980) "Sex ties and status attainment," unpublished paper, Temple University Department of Sociology.

Euske, N. A. and Roberts, K. H. (1987) "Evolving perspectives in organisation theory: Communication implications," in *Handbook of Organisational Communication: An Interdisciplinary Perspective*, F. M. Jablin, L. L. Putnam, K. H. Roberts and L. W. Porter (eds.), Sage, London, 41–69.

Fagan, B. (1995) *Ancient North America*. Thames & Hudson, London.

Featherstone, M. (1988) "In pursuit of the postmodern: An introduction," *Theory Culture and Society*, **5**(2/3), 195–215.

Fonseca, I. (1996) *Bury Me Standing*. Vintage, London.

Fulk, J. and Boyd, B. (1991) "Emerging theories of communication in organisations," *Journal of Management*, **17**(2), 407–446.

Garret, L. (1994) *The Coming Plague*. Penguin, London.

Gill, S. D. (1982) *Beyond the Primitive: The Religions of Nonliterate Peoples*. Prentice Hall, Englewood Cliffs, New Jersey.

Granoveter, M. S. (1973) "The strength of weak-ties," *American Journal of Sociology*, **78**, 1360–1380.

Harvey, D. (1990) *The Condition of Postmodernity*. Blackwell, Oxford.

Heckscher, C. (1994) "Defining the post-bureaucratic type," in *The Post-Bureaucratic Organisation*, C. Heckscher and A. Donnellon (eds.), Sage, London, 14–62.

Johnson, G. and Scholes, K. (1997) *Exploring Corporate Strategy*. Prentice Hall, London.

Josephy, A. M. (1994) *500 Nations: An Illustrated History of North American Indians*. Alfred A. Knopf, New York.

Levi-Strauss, C. (1961) *Tristes Tropiques*. (Translated J. Russell), Atheneum, New York.

Lyotard, J. F. (1984) *The Post Modern Condition*. (Translated G. Bennington and B. Massumi), University of Minneapolis Press, Minneapolis.

McAdam, D. (1983) "Tactical innovation and the pace of insurgency," *American Sociological Review*, **48**, 735–753.

Malinowski, B. (1922) *Argonauts of the Western Pacific*. Dutton, New York.

Marcus, G. E. and Fischer, M. J. (1986) *Anthropology as Cultural Critique*. University of Chicago Press, London.

Mead, M. (1949) *Coming of Age in Samoa*. Mentor Books, New York.

Newton, T. (1996) "Postmodernism and Action," *Organization*, **3**(1), 7–29.

Nohria, N. and Berkley, J. D. (1994) "The virtual organisation," in *The Post-Bureaucratic Organisation*, C. Heckscher and A. Donnellon (eds.), Sage, London, 108–128.

Nonaka, I. and Takeuchi, H. (1995) *The Knowledge-Creating Company: How Japanese Companies Create the Dynamics of Innovation*. Oxford University Press, New York.

Pease, T. C. (1965) *The Leveller Movement*. Oxford University Press, Oxford.

Phelan, S. E. (1995) "From chaos to complexity in strategic planning," paper presented at the *55th Annual Meeting of the Academy of Management*, Vancouver.

Pool, I. (1980) "Comment on Mark Granovetter's the strength of weak-ties: A network theory revisited," paper presented at the *International Communications Association Conference*, Acapulco.

Popper, K. R. (1945) *The Open Society and Its Enemies*. Routledge & Kegan Paul, London.

Radhakrishnan, R. (1994) "Postmodernism and the rest of the work," *Organization*, **1**(2), 305–340.

Scott, J. (1991) *Social Network Analysis: A Handbook*. Sage, London.

Simons, C. and Naylor-Stables, B. (1997) *Effective Communication for Managers*. Cassel, London.

Stacey, R. D. (1993). *Strategic Management and Organisational Dynamics*. Pitman, London.

Steward, F. and Conway, S. (1998) "Situating discourse in environmental innovation networks," *Organization*, **5**(4), 479–502.

Tarrow, S. (1995) "Cycles of collective action," in *Repertoires and Cyclés of Collective Action*, M. Traugott (ed.), Duke University Press, New York.

Tilley, C. (ed.) (1979) *An Urban World*. Little Brown, Boston, MA.

Vriend, N. J. (1994) "Self-organized markets in a decentralized economy," *Working Paper*, No. 94-03-013, Santa Fe Institute, CA.

Wellman, B. and Leighton, B. (1979) "Networks, neighbourhoods and communities," *Urban Affairs Quarterly*, **15**, 363–390.

Wuthnow, R. (1989) *Communities of Discourse*. Harvard University Press, Cambridge, MA.

Chapter 11

Realising the Potential of the Network Perspective in Researching Social Interaction and Innovation

Steve Conway*, Oswald Jones and Fred Steward

The Rise of the Network Perspective: Simply a Fashion or Fad?

Even a cursory review of the various business and management sub-disciplines in the post-war period reveals the ebb and flow of a whole series of management and organisational innovations: the 1980s heralded the rise of 'Quality Circles', 'Portfolio Management', MBWA ('Management by Wandering Around'), and the focus on 'Excellence', to name but a few; in the 1990s the fashionable themes shifted to 'Downsizing', 'TQM' (Total Quality Management), 'BPR' (Business Process Engineering), the 'Learning Organisation', and the 'Virtual Organisation'. Carter and Conway (2000:61) are sceptical about the motivation and utility surrounding the packaging and re-packaging of such managerial and organisational panaceas:

*E-mail: s.conway@aston.ac.uk

"The ability of the American academic-consulting complex (Grey and Mitev, 1995) to generate seductive panaceas for organisational action is not in doubt. Recent years have seen the quality movement (see Oakland, 1993), the excellence tradition (Peters and Waterman, 1982) and Business Process Re-engineering (Hammer and Champy, 1993) become part of the corporate lexicon, a feature of the organisational world... All of these movements are characterised by an attention to promotion, in which, grand claims are made, imploring individual managers and organisations to adopt and enrol within a given technique. In short, new managerial ideas are aggressively marketed, promising a great deal to adopting organisations (see Egan, 1995; Wilson, 1992); they are at once rhetoric and image intensive (see Alvesson, 1998)."

The 1990s have also seen the rapid rise in interest and reference to inter- and intra-organisational networks; journals and conferences in a broad range of business and management sub-disciplines from innovation studies, marketing, to public sector management, are littered with articles and subject streams concerning networks and networking behaviour. To what extent is this simply a passing fashion or fad? Does the network perspective have something useful and distinctive to offer academics, managers, and policy-makers? Or, in the words of Egan (1995), does the network approach amount to "a bauble or a breakthrough" in developing an understanding of organisations and organisational activity? Furthermore, to what extent does this simply reflect a change in ontology and epistemology, rather than a real change in the way organisations organise and operate?

Arguably, this decade has perhaps seen an over-use of the network perspective in studies of organisations; an over-use that has diluted its power and legitimacy in the academic world as having something 'original' and 'useful' to say about organisations and organising.

Nevertheless, this dramatic rise in the interest in organisational networks has brought the concept into the main-stream of academic, management, and policy-making discourse. Indeed, during the 1990s the network approach has successfully filtered through from the academic world into the policy arena (for example, see references to the importance of networking in the UK Government's White Paper on Innovation: 'Realising Our Potential' (1993); and numerous initiatives under the SPRINT programme, such as MINT — Managing the Integration of New Technologies); consultancies (for example, Coopers & Lybrand, 1994); and the business world. Within and through such diverse institutions, network building and networking activity have frequently been espoused as the panacea to the innovativeness of organisations, the regeneration of industrial regions, the encouragement of emerging industrial sectors, and for the revitalisation of the small firm sector. Bessant (1995:268) argues that "networks are a comparatively new addition to the policy-maker's tool-box, and there is considerable excitement about their potential". However, Bessant also suggests that as a consequence of this, there exists a risk that networks will be used widely and not always appropriately.

The utility of the network perspective is at least partially derived from the ease with which the concept can be expressed and applied. It is at once a concept and framework whose applicability is immediately recognisable by practitioners, whilst its academic pedigree has been firmly established, particularly within the field of anthropology where it has been widely and successfully applied in the form of social network analysis (SNA) since the 1960s. The flexibility of the network perspective, both in relation to the manner in which it may be applied (see Chapter 1), and the subject matter it may be applied to, from the diffusion of AIDS (Klovdahl, 1985) to the diffusion of innovation (Rogers and Shoemaker, 1971), has also been of great import to its success; it is a perspective that can throw light on phenomena at any unit of analysis, such as at the level of the individual (for example, the social organisation within firms (Allen,

1970) and within academic specialities (Crane 1972)), and at the level of the organisation (for example, revealing the structure of industrial sectors by identifying organisational networks (Hagedoorn and Schakenraad, 1992)). Furthermore, for Bianchi and Bellini (1991:489) "historic experience has taught us the need for a systemic approach to industrial organisation" and that "the 'network' is a stylized concept which we can use both as an analytical tool to understand economic reality and as a reference for political action in order to modify that reality". That is, the network concept is a useful framework for evaluating the structure and operation of existing networks, and for highlighting factors that might improve their performance.

At the level of the organisation, evidence suggests that networking activity, particularly within high technology sectors such as tele-communications, computing, and bio-technology, has been increasing rapidly since the 1980s (Hagedoorn and Schakenraad, 1993; Chen, 1997). In this sense, the network approach encapsulates an emerging phenomena rather than simply representing a change in ontology and epistemology. At the level of the individual, evidence suggests that extensive personal networks and networking have been important features of the creative and innovation process throughout the post-war period (for example, Menzel, 1962; Price and Beaver, 1966; Crane, 1972; Allen, 1970); here the re-emergence of the network perspective is revealing an existing phenomona, and thus represents a change in ontology and epistemology. It would seem then, that the network both as a perspective and a phenomena is more than simply a passing fad or fashion.

The 'Taken-for-Granteds' of the Network Perspective in Innovation Studies

As noted earlier, the 'network' as a metaphor is powerful way of viewing organisations; it changes the imagery from a focus on pairs of dyadic relationships to one of "constellations, wheels, and systems of relationships" (Auster, 1990:65) and of 'webs' of group affiliations (Simmel, 1955). This perspective is important, since as DeBresson and Amesse (1991:364) argue, "interactions between firms...are iterative and broad in content, time and space, [and] what matters is the complete set of relationships". However, Tichy *et al.* (1979:507) contend that:

> "Such a model of organising, if it is to move beyond the metaphorical stage, requires a coherent framework and accompanying methods of analysis that are capable of capturing both prescribed and emergent processes."

Yet, one of the most striking features of many of the articles concerning innovation and technology networks is the general lack of explicitness. This lack of explicitness can be seen in relation to the nature and boundary of the network under investigation, the nomenclature employed, and to the features of the network that are being revealed: in particular, the term 'network' is expected to speak for itself. However, the term 'network' means different things to different people, to the extent that there is little common ground beyond the metaphor. For example, is the ego-centred set of dyadic relationships of a given actor, a network, when it makes no reference to how the ego-centred network is embedded within the broader network?

The lack of explicitness in many network papers in the innovation and technology field would appear to stem from the fact that the dominant starting point is the imagery provided by the network metaphor. Some consider this to be 'imagery without technique' (Shrum and Mullins, 1988). Those studies and papers whose starting point is the social network literature, for example, draw from a long and rich tradition in which a great deal of focus has been placed upon creating a coherent framework for the systematic analysis of networks; thus, the social network literature explicates the various approaches to abstracting 'partial networks' from the 'total network' (see Table 3 in Chapter 3) and is thus more explicit about the rules of inclusion or exclusion of the actors, linkages, and flows under investigation. While the social network literature has much to offer innovation studies, Rogers (1987:14) warns that "far too much, I fear, we admire mathematical elegance in our network tools and tool-makers, while largely ignoring what useful objects we can dig up with these tools". Indeed, Wellman (1983:156) argues that network analysis should be viewed "as a broad intellectual approach, and not as a narrow set of methods". Nevertheless, the greater explicitness in the social network literature, the development of tools and techniques to systematise data collection, data analysis, and data presentation, as well as the emergence of some useful network theory, are important contributions that are often overlooked in 'looser' applications of the network approach. As the authors have argued in earlier publications, perhaps the adoption of a middle-ground (as illustrated in Fig. 1) between the extremes of the mathematical and metaphorical orientations may be a way forward (Conway, 1997a; Jones *et al.*, 1998). A graphical orientation has the potential to both amplify the imagery of the network metaphor whilst embracing a more systematic and explicit approach to collecting, analysing and presenting relational data

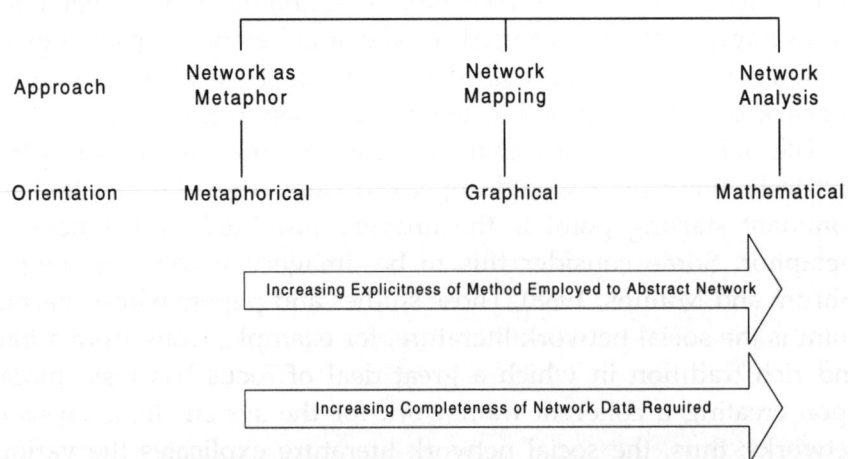

Fig. 1. Alternative approaches and orientations in studying networks (adapted Conway, 1997a; Jones *et al.*, 1998).

(Conway, 1997a; Conway and Steward, 1998a; 1998b; Steward and Conway, 1998). The potential of this approach is demonstrated in the chapters by Conway (on mapping informal networks), Steward (on mapping innovation networks and risk arenas, and the overlap of these two networks in the shaping of technology), and Jones and Beckinsale (on mapping interactions during a sample of innovation projects), and is one possible way in which innovation studies may realise greater utility from the adoption of a network perspective.

One of the ways in which the broader social network literature can usefully inform innovation network studies is through the efforts that have been placed in eludicating the key dimensions of both the network itself, and of the individual components that make up the network: i.e., actors, links, and flows (see Fig. 2). This literature provides a useful set of characteristics for comparing the morphology

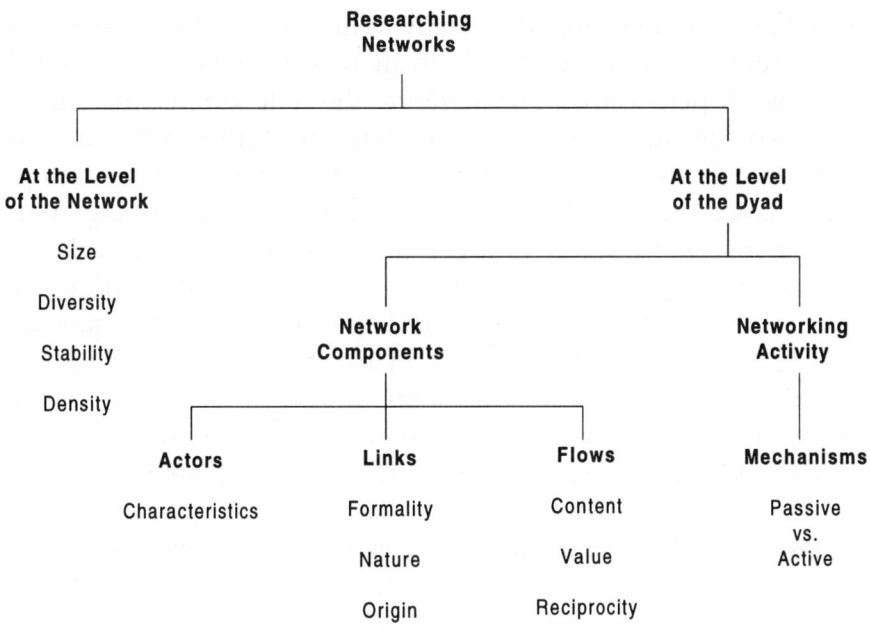

Fig. 2. Deconstructing the network and its components.

of networks over time, or between the networks of different firms or industrial sectors at a single point in time. The deconstruction of the network into its component parts, and of these component parts into a set of key characteristics, is also an important contribution for the systematic collection of data.

Issues Relating to Existing Network Research

While the network perspective has been widely adopted in a broad range of disciplines, such as economics, anthropology, social psychology and sociology, to study all manner of phenomena, there are few articles that take stock of the progress and gaps in network research. Nevertheless, such reflection is important if the network approach is to realise its potential; comments by Rogers (1987) and Salancik (1995) are particularly useful in this respect. Although Rogers and Salancik are referring specifically to the social network literature, their comments are generalisable to those studies broadly adopting the network perspective. Furthermore, the following comments are not aimed specifically at innovation network studies, although they have particular resonance with such research. In providing a critique of the network approach, Salancik (1995:345) starts by asking "what role a network perspective might uniquely play beyond acknow-ledging and highlighting the importance of social relations for organisational and inter-organisational affairs." This is a key question, since, as Salancik (1995) points out, many of the findings of network studies could have been obtained without the use of existing network theory and tools. By raising the issue of the nature of the uniqueness of the contribution provided by the network perspective, one is drawn into recognising both the existing weaknesses and potential of the network approach. The weaknesses in existing social network research include: (1) an under-emphasis on data-gathering; (2) a lack of focus on the dynamic of the network and the flow of content through the network; (3) an under-emphasis on theory building; and (4) a lack of explicitness in sampling and in the setting of the boundary of the investigation. These issues will now be dealt with in turn.

An Under-Emphasis on Data-Gathering

With reference to social network analysis, Rogers (1987:17) argues that "without good data, network analysis is worthless"; by 'good data' it would appear that Rogers is referring to the need for 'complete' data with respect to the actors and relationships within a network under investigation. This is particularly important where the focus is socio-centred and mathematical social network techniques are applied to the data to reveal network characteristics such as density and connectedness (this is illustrated in Fig. 1). However, the reference to the need for 'good data' also highlights a general under-emphasis on data-gathering in network studies, whether or not the researcher is adopting a mathematical orientation. It is not uncommon in the innovation studies literature for network data to be fairly shallow and superficial. In particular, more emphasis needs to be placed on gathering data on the nature of relationships and the various flows through these linkages in the network. There is also a paucity of longitudinal data on networks and their component parts. These more specific areas of data gathering will now be explored further.

Short-Changing Network Message Content and Time

Alba (1982) argues that network analysts have focused mainly on the form of networks, while largely ignoring the content of the information that flows through network links. Rogers (1987:19) suggests that "perhaps a major reason for ignoring network content is due to our overwhelming dependence in network data-gathering upon sociometric, who-to-whom questions". This concern is supported by Salancik (1995:346) who believes that "there is a danger in network analysis of not seeing the trees for the forest. Interactions, the building blocks of networks are too easily taken as givens". Innovation studies would certainly benefit from such a focus since key areas of interest include the management and sharing of tacit knowledge (for example, Nonaka and Takeuchi, 1995), and the sourcing of ideas and inputs in successful innovation (for example, Conway, 1995; Conway and Steward, 1998).

In addition, Rogers (1987:19) argues that "past network research has mainly been cross-sectional in nature, thus ignoring time as a variable". Salancik (1995:348) sees this as problematic: "A network analysis taken in a snapshot of time might miss the organising that is going on and the stable system that eventually evolves". However, a few network scholars have effectively brought time into their studies, including Klovdahl's (1985) investigation of the diffusion of AIDS, and more specific to the innovation literature, the work of Hagedoorn and Schakenraad (1992), concerning the trends in strategic alliances and joint ventures between leading firms in a number of information technology sectors. Again this is an area that is beginning to be addressed by a number of researchers at Aston Business School, for example, see the chapters by Conway, and Jones and Beckinsale.

Theory-Less Orientated Research

Rogers (1987:14) argues that "while we can easily point to numerous important network data-analysis contributions, it is much more difficult to identify really significant theoretical advances". Indeed, this view is supported by Burt (1980:134) who contends that "the lack of network theory seems to me to be the most serious impediment to the realization of the potential value of network models in empirical research". Granovetter (1979) has called this the theory-gap in network studies. Salancik (1995:348) is more pointed in his criticism:

"To be productive in understanding organisations, network analysts will need to become more theoretical about the things that they study. When questions about network effects are asked, the source of the questions are usually other theories. Thus many questions about interacting organisations are framed from resource dependency theory. Many questions about diffusion of practice or attitude within or between organisations are framed from general theories of social influence or social comparison. Network analysts often don't ask how their perspective addresses a theoretical problem, but how network analysis can be used to look into a problem area… Instead of capitalizsing on opportunities for applying the craft, network analysts will surely advance our under-standing of organisation better by constructing network theories about organisation…a network theory of organisation should propose how structures of interactions enable coordinated interaction to achieve collective and individual interests."

One such important network theory is the 'strength-of-weak-ties' (Granovetter, 1973); Rogers, (1987:14–15) recognises both the attractiveness "of its seemingly paradoxical nature" and the fact that network scholars altered their methods of measuring networks as a consequence. Another important contribution to network theory, is the related concept of 'structural holes' (Burt, 1992). Both of these network theories are important to developing our understanding of innovation and entrepreneurship (Conway, 1997b). It is only through such theoretical contributions that the network perspective will be able to develop a distinctive contribution to our understanding of organisations. In this regard, Krackhardt (1995:350) offers the following advice:

"A good theory of organisations builds on existing literature and yet provides new insight into otherwise confusing or contradictory phenomena. It must be falsifiable but should be accompanied by solid empirical support for its predictions.

It should be clear and sensible, but also interesting and not trivial. The theory is better if it can be generalised to a large set of phenomena, applying to a range of units of analysis, such as the individual, the organisation, and perhaps even larger social entities. If a theory satisfies all of these conditions, then it would also be nice if it stipulated practical implications for organisational members."

Making Explicit the Boundary of Analysis

Establishing the boundary of the network under investigation is one of the fundamental issues that need to be addressed when conducting research employing the network perspective. Yet despite this, few published studies are explicit in specifying the researchers rules of inclusion. Laumann *et al.* (1983:18) argue that "the problem of boundary definition should be given conscious attention" and that "care must be given to specifying the rules of inclusion" in relation to both "the selection of actors or nodes…and to the choice of types of social relationships to be studied". However, a key problem in network sampling arises from the difficulty in specifying the boundary of the network. Barnes (1979:416) argues that "networks are interesting but difficult to study since real-world networks lack convenient natural boundaries". In her study of the networks of scientists ('invisible colleges') Crane (1972:14) saw boundary setting as problematic, contending that "the amorphous character of [scientific] research areas complicates the problems of defining the membership of the social circles". However, the network researcher must also be wary of natural boundaries, since as Alba (1982:43) argues, "natural boundaries may at times prove artificial, insofar as individuals within the boundaries may be linked through others outside of them". Indeed, the network researcher has to be particularly wary in setting the boundary in investigations of innovative activity given the importance of boundary-spanning to the innovation process. This point is highlighted by the auther in Chapter 3:

"A key characteristic of informal and social organisation is their tendency to span organisational boundaries: team boundaries, functional boundaries, and even the organisational boundary itself. Such boundary spanning interaction is the essence of the *interactive model* of innovation (Rothwell and Zegveld, 1985). The interactive model places great emphasis on the ability of innovative organisations to manage relationships across interfaces, both within the firm (between project groups, functional departments, and divisions), and externally (within and across industrial sectors, geographical regions, and nations)."

For Fombrun (1982:288) the solution to the boundary-setting problem should be based on the objectives of the research, arguing that "if there is no agreed boundary to an inter-organisational network, the choice of the boundary should reflect the purposes of the researcher and the research hypotheses of the study". Mitchell (1969:40) supports this view, arguing that:

"Clearly some limit must be put on the number of links to be taken as definitive for any specific network, otherwise it would become co-extensive with the total network. This difficulty is resolved by fixing the boundary of the network in relation to the social situation being analyzed... There can be no general rule."

Nevertheless, researchers should be aware that the approach through which boundaries are drawn up is a critical step in the research process, since it creates the sample of linkages that are examined (Auster, 1990). Laumann *et al.* (1983) also note that carelessness, in what they term system specification, can distort the overall configuration of the network. With this in mind, Fombrun (1982:288) warns that the "conclusions drawn from the study must be carefully scrutinized for the possibility of alternative explanations grounded in the effects of the untapped networks".

Concluding Comments

While it is true to say that the network perspective has been particularly popular among academics, policy-makers, and managers alike, during the 1990s, it has been argued that networking is likely to have a more enduring impact than other management fashions of recent years. Importantly, both the perspective and the phenomena are seen to have great utility among a broad range of constituents. In part, this is due to the ease with which the concept can be expressed and applied by both practitioners and academics. The network perspective also offers a great deal of flexibility in the manner in which it may be applied and the subject matter or phenomena it may be applied to. However, it has also been argued that there are too many 'taken-for-granteds' in the application of the network approach within innovation studies and that there needs to be greater explicitness and depth in its usage by academics if the field is to gain the full potential from the perspective. In this respect, there is much to learn from the broader social network literature, particularly from anthropology and sociology. In order that the network perspective matures within innovation studies in such a way as to provide a distinctive and useful contribution, it has been argued that various needs require attention: (1) the need for 'deeper' data-gathering that goes beyond the investigation of structure; (2) a greater focus on the dynamic of the network and the flow of content through the network; (3) the need for theory building; and (4) the need for greater explicitness in sampling and in the setting of the boundary of the investigation. The various chapters in this book begin to address these issues, but much work still needs to be undertaken if the full potential of the network perspective is to be realised within innovation studies. The greatest challenge, perhaps, is the development of useful and robust theory.

References

Alba, R. (1982) "Taking stock of network analysis: A decade's results," in *Research in the Sociology of Organizations: A Research Annual*, Vol. 1, S. Bacharach (ed.), JAI Press, Connecticut, 39–74.

Allen, T. (1970) "Communication networks in R&D laboratories," *R&D Management*, **1**(1), 14–21.

Auster, E. (1990) "The interorganizational environment: Network theory, tools, and applications," in *Technology Transfer: A Communication Perspective*, F. Williams and D. Gibson (eds.), Sage Publications, 63–89.

Baaijens, J. (1998) "The social structure of innovation and entrepreneurship," paper presented at the *10th International SASE Conference*, Vienna.

Barnes, J. (1979) "Network analysis: Orienting notion, rigorous technique, or substantive field of study," in *Perspectives on Social Research*, P. Holland and S. Leinhardt (eds.), Academic Press, New York, 403–423.

Bessant, J. (1995) "Networking as a mechanism for enabling organisational innovations," in *Europe's Next Step: Organisational Innovation, Competition and Employment*, L. Andreasen, B. Coriat and D. Friso (eds.), Frank Cass & Co., London, 253–270.

Bianchi, P. and Bellini, N. (1991) "Public policies for local networks of innovators," *Research Policy*, **20**(5), 487–497.

Burt, R. (1980) "Models of network structure," in *Annual Review of Sociology*, Vol. 6, A. Inkeles (ed.), 79–141.

Burt, R. (1992) *Structural Holes: The Social Structure of Competition*. Harvard University Press, Cambridge, MA.

Carter, C. and Conway, S. (2000) "A manifesto for corporate myopia: A cautionary note on shareholder value techniques," in *Value-Based Management*, G. Arnold and M. Davies (eds.), Wiley, 61–80.

Chen, S. (1997) "A new paradigm for knowledge-based competition: Building an industry through knowledge sharing," *Technology Analysis and Strategic Management*, **9**(4), 437–452.

Conway, S. (1995) "Informal boundary-spanning networks in successful technological innovation," *Technology Analysis & Strategic Management*, 7(3), 327–342.

Conway, S. (1997a) "Focal innovation action-sets: A methodological approach for mapping innovation networks," *Research Paper Series*, No. RP9702, Aston Business School Research Institute, Birmingham.

Conway, S. (1997b) "Strategic personal links in successful innovation: Link-pins, bridges, and liaisons," *Creativity and Innovation Management*, 6(4), 226–233.

Conway, S. and Steward, F. (1998a) "Mapping innovation networks," *International Journal of Innovation Management*, 2(2), 165–196.

Conway, S. and Steward, F. (1998b) "Networks and interfaces in environmental innovation: A comparative study in the UK and Germany," *Journal of High Technology Management Research*, 9(2), 239–253.

Coopers & Lybrand (1994) *Good Practice in Managing Transnational Technology Transfer Networks: 10 Years of Experience in the SPRINT Programme.* Vol. 1, subject papers.

Crane, D. (1972) *Invisible Colleges: Diffusion of Knowledge in Scientific Communities.* University of Chicago Press, Chicago.

DeBresson, C. and Amesse, F. (1991) "Networks of innovators: A review and introduction to the issue," *Research Policy*, 20(5), 363–379.

Egan, C. (1995) *Creating Organizational Advantage.* Butterworth Heinemann, London.

Fombrun, C. (1982) "Strategies for network research in organisations," *Academy of Management Review*, 7(2), 280–291.

Granovetter, M. (1973) "The strength of weak ties," *American Journal of Sociology*, 78(6), 1360–1380.

Granovetter, M. (1979) "The theory-gap in social network analysis," in *Perspectives on Social Research*, P. Holland and S. Leinhardt (eds.), Academic Press, New York, 501–518.

Hagedoorn, J. and Schakenraad, J. (1992) "Leading companies and networks of strategic alliances in information technologies," *Research Policy*, **21**(2), 163–190.

Jones, O., Conway, S. and Steward, F. (1998) "Introduction: Social interaction and innovation networks," *International Journal of Innovation Management*, **2**(2), 123–136.

Klovdahl, A. (1985) "Social networks and the spread of infectious diseases: The AIDS example," *Social Science of Medicine*, **12**, 1203–1216.

Krackhardt, D. (1995) "Review symposium: Structural holes," *Administrative Science Quarterly*, **40**, 350–354.

Laumann, E., Marsden, P. and Prensky, D. (1983) "The boundary specification problem in network analysis," in *Applied Network Analysis: A Methodological Introduction*, R. Burt and M. Minor (eds.), Sage, Beverly Hills, 18–34.

Menzel, H. (1962) "Planned and unplanned scientific communication," in *The Sociology of Science*, B. Barber and W. Hirsch (eds.), Free Press, New York, 417–441.

Mitchell, J. (1969) *Social Networks in Urban Situations*. Manchester University Press.

Nonaka, I. and Takeuchi, H. (1995) *The Knowledge Creating Company*. Oxford University Press.

Price, D. and Beaver, D. (1966) "Collaboration in an invisible college," *American Psychologist*, **21**, 1011–1018.

Rogers, E. (1987) "Progress, problems and prospects for network research: Investigating relationships in the age of electronic communication," paper presented at the *VII Sunbelt Social Networks Conference*, Florida, 12–15 February.

Rogers, E. and Shoemaker, F. (1971) *Communication of Innovations: A Cross Cultural Approach*. Free Press, New York.

Salancik, G. (1995) "Wanted: A good network theory of organization," *Administrative Science Quarterly*, **40**, 345–349.

Simmel, G. (1955) *Conflict and the Web of Group-Affiliation*. (Translated by K. Wolff and R. Bendix), Free Press, New York.

Shrum, W. and Mullins, N. (1988) "Network analysis in the study of science and technology," in *Handbook of Quantitative Studies of Science and Technology*, A. van Rann (ed.), North-Holland, Amsterdam.

Steward, F. and Conway, S. (1998) "Situating discourse in environmental innovation networks," *Organization*, **5**(4), 483–506.

Tichy, N., Tushman, M. and Fombrun, C. (1979) "Social network analysis for organisations," *Academy of Management Review*, **4**(4), 507–519.

Wellman, B. (1983) "Network analysis: Some basic principles," in *Sociological Theory*, R. Collins (ed.), Jossey-Bass, San Francisco.

Index